FLEXIBLE TEXT SEARCHING

FLEXIBLE TEXT SEARCHING

PANAGIOTIS D. MICHAILIDIS
AND
KONSTANTINOS G. MARGARITIS

Nova Science Publishers, Inc.
New York

© 2008 by Nova Science Publishers, Inc.

All rights reserved. No part of this book may be reproduced, stored in a retrieval system or transmitted in any form or by any means: electronic, electrostatic, magnetic, tape, mechanical photocopying, recording or otherwise without the written permission of the Publisher.

For permission to use material from this book please contact us:
Telephone 631-231-7269; Fax 631-231-8175
Web Site: http://www.novapublishers.com

NOTICE TO THE READER

The Publisher has taken reasonable care in the preparation of this book, but makes no expressed or implied warranty of any kind and assumes no responsibility for any errors or omissions. No liability is assumed for incidental or consequential damages in connection with or arising out of information contained in this book. The Publisher shall not be liable for any special, consequential, or exemplary damages resulting, in whole or in part, from the readers' use of, or reliance upon, this material.

Independent verification should be sought for any data, advice or recommendations contained in this book. In addition, no responsibility is assumed by the publisher for any injury and/or damage to persons or property arising from any methods, products, instructions, ideas or otherwise contained in this publication.

This publication is designed to provide accurate and authoritative information with regard to the subject matter cover herein. It is sold with the clear understanding that the Publisher is not engaged in rendering legal or any other professional services. If legal, medical or any other expert assistance is required, the services of a competent person should be sought. FROM A DECLARATION OF PARTICIPANTS JOINTLY ADOPTED BY A COMMITTEE OF THE AMERICAN BAR ASSOCIATION AND A COMMITTEE OF PUBLISHERS.

Library of Congress Cataloging-in-Publication Data
Michailidis, Panagiotis D.
　　Flexible text searching / Panagiotis D. Michailidis and Konstantinos G. Margaritis.
　　　p. cm.
　　ISBN 978-1-60692-261-3 (pbk.)
1. Computer algorithms. 2. Database searching. 3. Electronic data processing–Distributed processing. 4. Parallel processing (Electronic computers) I. Margaritis, Konstantinos G. II. Title.
　　QA76.9.A43M54 2009
　　025.5'24–dc22
　　　　　　　　　　　　2008043937

Published by Nova Science Publishers, Inc. ✤ *New York*

Contents

Preface	**vii**
1 Introduction	**1**
2 Exact String Matching Algorithms	**3**
2.1. Problem Defintion	3
2.2. Algorithms for Exact String Matching	3
2.2.1. Classical Approach	3
2.2.2. Suffix Automata Approach	4
2.2.3. Bit Parallelism Approach	5
2.2.4. Hashing Approach	5
2.3. Experimental Map	5
3 Approximate String Matching Algorithms	**7**
3.1. String Matching with k Mismatches	7
3.1.1. Problem Definition	7
3.1.2. Algorithms for String Matching with k Mismatches	7
3.1.3. Experimental Map	9
3.2. String Matching with k Differences	9
3.2.1. Problem Definition	9
3.2.2. Algorithms for String Matching with k Differences	10
3.2.3. Experimental Map	13
4 Parallel and Distributed Computing	**15**
4.1. Parallel Computer Architectures	15
4.1.1. General Purpose Parallel Computers	16
4.1.2. Special Purpose Parallel Hardware	16
4.2. Towards Low Cost Parallel Computing and Motivations	17
4.3. Architecture of a Cluster Computer	18
4.4. Programming Environment and Tools	19
4.4.1. Threads	20
4.4.2. Message Passing	20
4.5. Parallel Programming Models	20
4.5.1. The Master-Worker Model	21
4.5.2. The Pipeline Model	21

	4.6. Mapping Algorithms to Array Processors	22
	4.6.1. Deriving Dependence Graph from Given Algorithms	22
	4.6.2. Mapping Dependence Graph onto Array Processors	22
5	**MPI Implementations of Exact and Approximate String Matching**	**25**
	5.1. The MPI Static Master-Worker Implementation	25
	5.2. The MPI Dynamic Master-Worker Implementation	26
	5.2.1. Dynamic Allocation of the Subtexts	26
	5.2.2. Dynamic Allocation of Text Pointers	27
	5.3. The MPI Hybrid Master-Worker Implmentation	27
	5.3.1. Text Distribution and Load Balancing	27
6	**Description of Algorithms for Implementation**	**29**
	6.1. Dynamic Programming Algorithms	29
	6.1.1. Preprocessing Phase	29
	6.1.2. Exact String Matching	30
	6.1.3. Approximate String Matching with k Mismatches	30
	6.1.4. Approximate String Matching with k Differences	31
	6.1.5. Approximate String Matching with k Differences Based on Myers Algorithm	31
	6.2. Nondeterministic Finite Automata (NFA) Algorithms	32
	6.2.1. Algorithm Based on the Rows of the Automaton	32
	6.2.2. Algorithm based on the Columns of the Automaton	35
	6.3. Extensions	36
	6.3.1. Limited Expressions	36
7	**Data Dependence Graphs for Approximate String Matching Algorithms**	**39**
	7.1. Dependence Graphs for the Dynamic Programming Algorithms	39
	7.2. Dependence Graphs for the NFA Algorithms	41
8	**Mapping Approximate String Matching Algorithms onto Processor Arrays**	**45**
	8.1. Processor Arrays for the Dynamic Programming Algorithms	45
	8.2. Processor Arrays for the NFA Algorithms	47
	8.3. An Implementation of the Preprocessing Phase	48
9	**A Unified Array Processor Architecture**	**51**
	9.1. Architecture of the Array	51
	9.2. Architecture of the Cells	53
10	**Comparison with Previous Hardware**	**57**
11	**Conclusion**	**59**
	References	**61**

Preface

The exact and approximate string matching problem is a common and often repeated task in information retrieval and bioinformatics. As current free textual databases are growing almost exponentially over time, the string matching problem is becoming more expensive in terms computational times. We believe that recent advances in parallel and distributed processing techniques are currently mature enough and can provide powerful computing means convenient for overcoming this string matching problem.

In this chapter we present a short survey for well known sequential exact and approximate string searching algorithms. Further, we propose four text searching implementations onto general purpose parallel computer like a cluster of heterogeneous workstations using MPI message passing library. The first three parallel implementations are based on the static and dynamic master-worker methods. Further, we propose a hybrid parallel implementation that combines the advantages of static and dynamic parallel methods in order to reduce the load imbalance and communication overhead. Moreover, we present linear processor array architectures for flexible exact and approximate string matching. These architectures are based on parallel realization of dynamic programming and non-deterministic finite automaton algorithms. The algorithms consist of two phases, i.e. preprocessing and searching. Then, starting from the data dependence graphs of the searching phase, parallel algorithms are derived, which can be realized directly onto special purpose processor array architectures for approximate string matching. Further, the preprocessing phase is also accommodated onto the same processor array designs. In addition, the proposed architectures support flexible patterns, i.e., patterns with a "don't care" symbol, patterns with a complement symbol and patterns with a class symbol. Finally, this chapter proposes a generic design of a programmable array processor architecture for a wide variety of approximate string matching algorithms to gain high performance at low cost. Further, we describe the architecture of the array and the architecture of the cell in more detail in order to efficiently implement for both the preprocessing and searching phases of most string matching algorithms.

Chapter 1

Introduction

The basic string searching problem can be defined as follows: Let a given alphabet (a finite sequence characters) Σ, a short pattern string $p = p_1 p_2 ... p_m$ of length m and a large text string $t = t_1 t_2 ... t_n$ of length n, where both the pattern and the text are sequences of characters from Σ with $m << n$. The string searching problem consists of finding one or more generally all the exact occurrences of a pattern p in a text t. This is one of the oldest and most pervasive problems in computer science.

The approximate string searching problem is a generalization of the exact string searching problem, which involves finding substrings of a text string close to a given pattern string. More specifically, the approximate string searching problem can be formally stated as follows: Let a given alphabet Σ, a short pattern string p of length m, a large text string t of length n with $m << n$, an integer $k \geq 0$ and a distance function d. This problem consists of finding all the substrings s of t such that $d(p,s) \leq k$.

The distance $d(p,s)$ between two strings p and s over an alphabet Σ is the cost of the minimum cost sequence of operations that needed to transform p into s. The cost of a sequence of operations is the sum of the costs of the individual operations. The cost of an operation is considered a positive real number.

In particular, in string searching applications the most interesting edit operations are: (a) changing one character to another single character (or a substitution), (b) deleting one character from the given string (or a deletion), and (c) inserting a single character into the given string (or an insertion).

There are several distance functions; two very well known functions are the Hamming distance and Levenshtein distance which are used in this chapter. The Hamming distance between two strings of equal length is defined as the number of positions with mismatching characters in the two strings. In other words, it allows only substitutions, which cost l. The approximate string searching problem with d being Hamming distance is called string searching with k mismatches. The Levenshtein or edit distance between two strings of not necessarily equal lengths, is the minimum number of character insertions, deletions and substitutions, which all cost l, required to transform the one string into the other. Algorithms for computing the edit distance between a pair of strings are presented in [59, 91, 94]. The approximate string searching problem with d being the Levenshtein or edit distance is called string searching with k differences (or sometimes string searching with k errors). Together the above two problems are called approximate string searching.

The solutions to the exact and approximate string searching problems differ if the algorithm has to be on-line (that is, the text is not known in advance) or off-line (the text can be preprocessed). In this chapter, we focus on on-line sequential algorithms for the these two problems. There are numerous algorithms for these problems, see for example the reviews of [62, 63, 71].

The exact and approximate string matching problem can be extended to include more flexible patterns, including patterns with a don't care symbol, patterns with a complement symbol and patterns with a class symbol. A don't care symbol can represent the matching with any single character, a complement symbol can represent the matching with all characters except the one that is complemented and a class symbol allows the matching with a subrange of characters. Some recent publications, which also provide surveys of previous work are [5, 10, 12, 26, 42, 51, 70, 96, 97]. Therein sequential and/or theoretical parallel algorithms are proposed for the solution of several aspects of the flexible approximate string matching problem.

Applications requiring some form of string matching can be found virtually everywhere. However, recent years have witnessed a dramatic increase of interest in sophisticated string matching problems like the approximate string matching problem, especially within the rapidly growing communities of information retrieval and computational biology. Not only these communities facing a drastic increase in the text sizes they have to manage, but they are demanding more and more powerful searches. As current free text collections are growing almost exponentially with the time, the string matching problem becomes impractical to use the fastest sequential algorithms on a conventional sequential computer system. To improve the performance of string searching on large text collections, we believe that recent advances in parallel and distributed processing is currently mature enough and can provide powerful computing means convenient for overcoming this string matching problem.

The goal of this chapter is to present sophisticated and intelligent methods by area of the parallel and distributed processing so that to speed up the searching on large text collections. The rest of this chapter is organized as follows: in the next section we briefly present the on-line sequential algorithms for the exact string matching problem. In section three we briefly describe the algorithms both for the k-mismatches and the k-differences problem. In section four we give a short background by area of the parallel and distributed computing. In section five we present four parallel methods of the flexible exact and approximate string matching algorithms on a distributed architecture general purpose such as a cluster of heterogeneous workstations using MPI library. In section six we describe the computational intensive flexible approximate string matching algorithms. In sections seven and eight we give implementations of computational intensive algorithms for flexible string matching problem onto architectures special purpose such as array processors. In section nine we propose an unified design of a programmable array processor architecture suitable for efficient execution of a class of flexible approximate string matching algorithms on a FPGA device to gain high performance at low cost. Finally, in the last section we present the conclusions of this chapter.

Chapter 2

Exact String Matching Algorithms

In this section we present the formal description of exact string matching problem as well as the basic on-line sequential algorithms. However, for the further details and the coding of the algorithms, the reader is referred to [61] and the original references. In general, an on-line string matching algorithm consists of two phases: the preprocessing phase in p and the search phase of p in t. During the preprocessing phase a data structure X is constructed, X is usually proportional to the length of the pattern and its details vary in different algorithms. The search phase uses the data structure X and it tries to quickly determine if the pattern occurs in the text. This phase is based on four different approaches including classical, suffix automata, bit-parallelism and hashing algorithms.

2.1. Problem Defintion

Let a given alphabet (a finite sequence characters) Σ, a short pattern string $p = p_1 p_2 ... p_m$ of length m and a large text string $t = t_1 t_2 ... t_n$ of length n, where both the pattern and the text are sequences of characters from Σ with $m << n$. The string matching problem consists of finding one or more generally all the exact occurrences of a pattern p in a text t.

2.2. Algorithms for Exact String Matching

2.2.1. Classical Approach

The classical string matching algorithms are based on character comparisons. The Brute-Force (in short, BF algorithm) algorithm, which is the simplest, performs character comparisons between a character in the text and a character in the pattern from left to right. In any case, after a mismatch or a complete match of the entire pattern it shifts exactly one position to the right. It requires no preprocessing phase and no extra space. The BF algorithm has $O(mn)$ worse-case time complexity. The average number of character comparisons is $n(1 + 1/(|\Sigma| - 1))$.

The Knuth-Morris-Pratt (in short, KMP) [45] algorithm, which was the first linear time string matching algorithm discovered, performs character comparisons from left to right. In case of mismatch it uses the knowledge of the previous characters that we have already examined in order to compute the next position of the pattern to use. In addition, this algorithm provides the advantage that the pointer in the text is never decremented. The preprocessing phase of the KMP algorithm requires $O(m)$ time and space. The searching phase needs $O(n)$ time in the worse and average case.

The next algorithm is Boyer-Moore (in short, BM) [3] algorithm, which is known to be very fast in practice, performs character comparisons between a character in the text and a character in the pattern from right to left. After a mismatch or a complete match of the entire pattern it uses two shift heuristics to shift the pattern to the right. These two heuristics are called the occurrence heuristic and the match heuristic. For the length of the shift is the maximum shift between the occurrence heuristic and the match heuristic. The details for two heuristics are referred to original paper [3]. These heuristics are preprocessed in $O(m+|\Sigma|)$ time and space. The searching phase of the BM algorithm needs $O(n+rm)$ time in the worse case, where r is the number of occurrences of the pattern in the text. Finally, the expected performance of the BM algorithm is sublinear requiring about n/m character comparisons on average.

The Boyer-Moore-Horspool (in short, BMH) [37] algorithm does not use the match heuristic. In case of mismatch or match of the pattern, the length of the shift is maximized by using only the occurrence heuristic for the text character corresponding to the rightmost pattern character (and not for the text character where the mismatch occurred). The preprocessing phase of the BMH algorithm requires $O(m+|\Sigma|)$ time and reduces the space requirements from $O(m+|\Sigma|)$ to $O(|\Sigma|)$. Finally, the searching phase requires $O(mn)$ time in the worse case but it can be proved that the average number of character comparisons is $n/|\Sigma|$.

The Quick Search (in short, QS) [87] algorithm of Sunday, performs character comparisons from left to right from the leftmost pattern character and in case of mismatch it computes the shift with the occurrence heuristic for the first text character after the last pattern character by the time of mismatch. The preprocessing and searching time of the QS algorithm are same as the BMH algorithm.

The Boyer-Moore-Smith (in short, BMS) [85] algorithm, noticed that computing the shift with the text character just next the rightmost text character gives sometimes shorter shift than using the rightmost text character. He advised then to take the maximum between the two values. The preprocessing phase of the BMS algorithm consists of $O(m+|\Sigma|)$ time and $O(|\Sigma|)$ space. Further, this algorithm has $O(mn)$ worse case time complexity.

The Turbo-BM (in short, TBM) [22] algorithm is an variant of the BM algorithm. It consists in remembering the substring of the text that matched a suffix of the pattern during the last character comparisons (and only if a good suffix shift has been performed). This method has two advantages: a) it is possible to jump over this substring and b) it can enable to perform a turbo shift. The details for the turbo shift is referred to original paper [22]. It can be shown that the number of character comparisons performed by the TBM algorithm is bounded by $2n$.

2.2.2. Suffix Automata Approach

This category uses the suffix automaton data structure (frequently called DAWG- for Deterministic Acyclic Word Graph) that recognizes all the suffixes of the pattern [23, 69].

The Reverse Factor (in short, RF) [22,55] algorithm, which performs the characters of the text from right to left using the smallest suffix automaton of the reverse pattern. The preprocessing phase of the RF algorithm requires linear time and space in the length of the pattern. The searching phase of RF algorithm has a quadratic worse-case time complexity but it is optimal on the average. It performs $O(nlogm/m)$ characters comparisons on the average.

2.2.3. Bit Parallelism Approach

Bit parallelism [4,5] uses the intrinsic parallelism of the bit manipulations inside computer words to perform many operations in parallel (whose number of bits in the computer word we denote w). This technique has became a general way to simulate simple nondeterministic finite automata (NFA) instead of converting them to deterministic. The main advantages of this approach are simplicity, flexibility and no buffering.

The basic idea of the first Shift-Or (in short, SO) [5] algorithm, is to represent the state of the search as a number, and each search step costs a small number of arithmetic and logical operations, provided that the numbers are large enough to represent all possible states of the search. Assuming that the pattern length is no longer than the computer word of the machine, the time complexity of the preprocessing phase is $O((m+|\Sigma|)\lceil m/w \rceil)$ using $O(m|\Sigma|)$ extra space. Finally, the time complexity of the searching phase is $O(n\lceil m/w \rceil)$ in the worse and average case, where $\lceil m/w \rceil$ is the time to compute a shift or other simple operation on numbers of m bits using a word size of w bits.

An new algorithm has appeared recently, called Backward Nondeterministic DAWG Matching (BNDM) [69]. This algorithm uses a nondeterministic suffix automaton that is simulated using bit parallelism. The preprocessing time for the BNDM algorithm is $O(m+|\Sigma|)$ for $m \leq w$ using $O(m|\Sigma|)$ extra space. The searching time is $O(mn)$ in the worse case and $O(nlogm/m)$ on average.

2.2.4. Hashing Approach

We introduce a different approach to string matching, the Karp-Rabin (in short, KR) [61] algorithm, which uses hashing techniques. Hashing provides a simple method to avoid a quadratic number of character comparisons in most practical situations. The main idea of the KR algorithm is to compute the signature or hashing function of each possible m-character substring in the text and check if it is equal to the signature function of the pattern. The preprocessing phase of the KR algorithm requires $O(m)$ time while the searching phase has $O(mn)$ worse case time complexity. Its expected number of character comparisons is $O(m+n)$.

2.3. Experimental Map

We present in this section a map of the efficiency of different string matching algorithms, showing zones where they are most efficient in practice. The experiments were performed on a $w = 32$ bits Ultra Sparc 1 running SunOs 2.5. Texts of 150 KB were randomly built, as were the patterns.

The map is shown in Figure 2.1. We observe that only QS, SO, BNDM and RF have a zone in the map since the others were too slow.

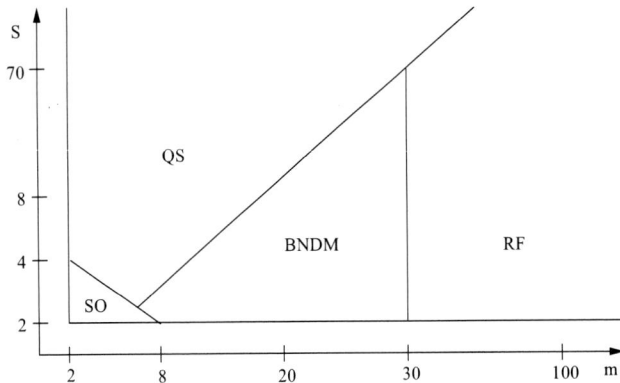

Figure 2.1. Map of experimental efficiency for different string matching algorithms.

The map shows clearly that the QS algorithm is efficient as the size of the alphabet and the pattern is increased. Further, the SO algorithm is fastest for short patterns and small alphabets. The BNDM algorithm is confined to a small zone for small alphabet sizes and moderate patterns. Finally, the RF algorithm is the best for long patterns and small/moderate alphabet sizes.

Chapter 3

Approximate String Matching Algorithms

In this section we give the formal description of string searching with k mismatches and string searching with k differences as well as the on-line sequential algorithms. However, for the further details and the coding of the algorithms, the reader is referred to the original references. In general, an on-line approximate string searching algorithm consists of two phases: the preprocessing phase in p and the searching phase of p in t. The preprocessing phase involves gathering of information about the pattern which can be used for a fast implementation of primitive operations in the searching phase or of constructing a finite automaton that recognizes all strings at a distance at most k from the pattern. The searching phase consists of scanning the text or the construction of a array in order to find all approximate occurrences of the pattern in the text. In general, the searching phase is based on four different approaches including dynamic programming/classical, deterministic finite automata, filtering, counting and bit-parallelism algorithms.

We start by reviewing the basic algorithms from each category for the string searching with k mismatches problem.

3.1. String Matching with k Mismatches

3.1.1. Problem Definition

Let a given alphabet Σ, a short pattern string $p = p_1 p_2 ... p_m$ of length m, a large text string $t = t_1 t_2 ... t_n$ of length n, in an alphabet Σ of size $|\Sigma|$, where $m, n > 0$ and $m << n$, and an integer maximum number of mismatches allowed $k \geq 0$, find all the text positions j such that in at most k positions in which p and t have different characters. We say that there is an approximate occurrence of p at position j of t.

3.1.2. Algorithms for String Matching with k Mismatches

Classical Approach

The classical string searching algorithms are based on character comparisons.

The Brute-Force algorithm (in short, BF algorithm), which is the simplest, performs character comparisons between the text substring and the complete pattern from left to right and to count the number of mismatches found. If more than k have been found, shifts exactly one position to the right. When we reach the end of the pattern we report an approximate occurrence. It requires no preprocessing phase. This algorithm has $O(mn)$ worst-case time complexity.

Landau-Vishkin algorithm (in short, LV) [49] developed the first efficient algorithm for this problem. Their approach is similar to the Knuth-Morris-Pratt algorithm [45] in that an array derived from preprocessing the pattern is employed as the text string is examined from left to right, and known information is exploited to reduce the number of character comparisons required. The preprocessing phase has $O(kmlogm)$ time and the searching phase has $O(kn)$ time, to the overall total. The extra space is required by the algorithm is $O(k(m+n))$. While it is fast, the space required is unacceptable for practical purposes. In our experiments we include the improved version of LV algorithm using a window of size $O(m^2)$ to process the text, instead of the $O(mn)$ array suggested in the original paper [49]. Therefore, this algorithm decreases the extra space to $O(km)$ which is acceptable in practice.

The next algorithm is Tarhio-Ukkonen algorithm (in short, TU) [88] which is based on the Boyer-Moore-Horspool (BMH) algorithm to exact string searching [37]. The preprocessing phase of TU algorithm has $O(m+k|\Sigma|)$ time and $O(k|\Sigma|)$ space. In searching phase of the TU algorithm needs $O(mn)$ time in the worst case. However, the expected running time is $O(kn((k/|\Sigma|)+1/(m-k)))$ for random strings.

Counting Approach

This approach does not use character comparisons like the classical algorithms, but it uses arithimetical operations i.e. it uses counters for every position of the text.

In this category we present only the Baeza-Yates-Perleberg algorithm (in short, BYP) [11] which is a very practical and simple solution to the string searching with k mismatches problem and whose performance is independent of k. This algorithm runs in $O(n)$ worst case time if all the characters in p are distinct and in $O(n+R)$ worst case time if there are identical characters in p, where R is the total number of ordered pairs of positions at which p and t match. Assuming the characters to be equiprobable, the average running time is $O((1+m/|\Sigma|)n)$, irrespective of the number of distinct pattern characters. Finally, the running time for preprocessing is $O(2m+|\Sigma|)$ and the space requirement for this algorithm is $O(m+|\Sigma|)$.

Bit-Parallelism Approach

This is another technique of common use in string searching [4]. It was first proposed in [5] for exact string searching problem. Therefore, this approach can be applied similar way to k mismatches problem.

From this category we include in our study the numerical algorithm Shift-Or (in short, SO) [5] which is introduced by Baeza et al.. This algorithm handles mismatches by essentially counting k of them with a $log_2 k$ size counter, but does not handle for the k differences problem i.e. deletions and insertions. Moreover, the bigger the number of bits needed to represent individual states, the smaller the length of patterns that are considered.

3.1.3. Experimental Map

We now present a map of the most efficient string matching algorithms with k mismatches. To give an idea of the areas where each algorithm dominates, Figure 3.1 shows the cases of English text and binary text. We must noted that the bit-parallelism algorithms which are occurred in the map correspond to a word size of $w = 32$ bits.

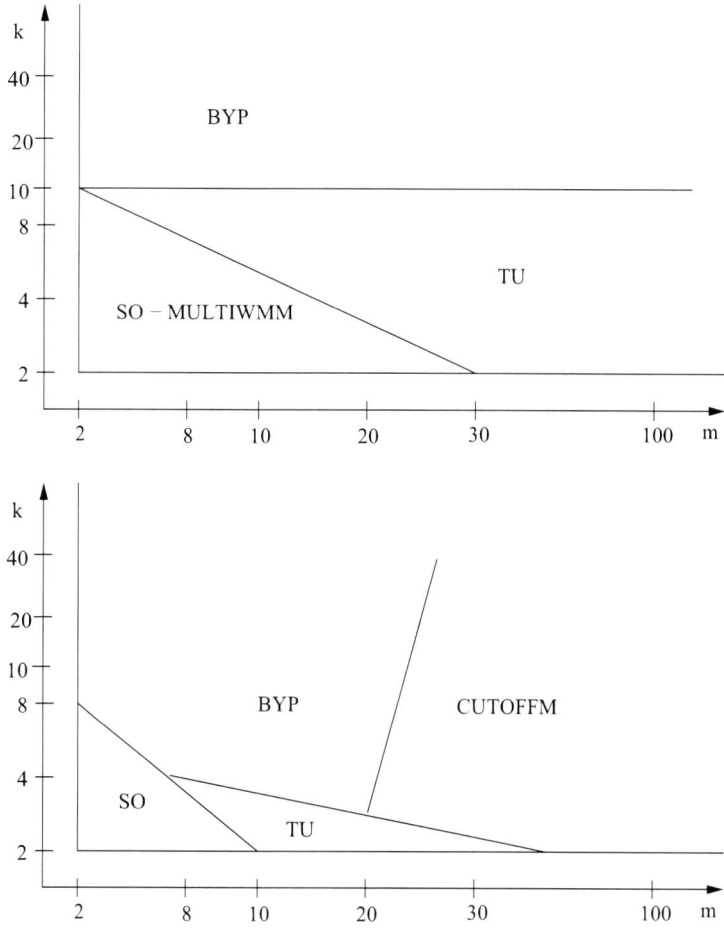

Figure 3.1. The areas where each string matching algorithm is best. English text is on top and binary text on the bottom.

3.2. String Matching with k Differences

3.2.1. Problem Definition

Let a given alphabet Σ, a short pattern string $p = p_1 p_2 ... p_m$ of length m, a large text string $t = t_1 t_2 ... t_n$ of length n, in a alphabet Σ of size $|\Sigma|$, where $m, n > 0$ and $m << n$, and an integer maximum number of differences allowed $k \geq 0$, find all the text positions j such that the edit distance (i.e. number of differences) between p and some substring of t ending an t_j is at most k. We say that there is an approximate occurrence of p at position j of t.

3.2.2. Algorithms for String Matching with k Differences

Dynamic Programming Approach

The dynamic programming approach is a classical solution which have been proposed independently by many researchers and mainly by the Wagner and Fischer [94] for computing the edit distance between two strings, the distances between longer and longer prefixes of the strings are successively evaluated from previous values until the final result is obtained. Later, Sellers [81] converts this classical solution into a search algorithm in order to find all approximate occurrences of p in the t. This algorithm has running time $O(mn)$ in the worst and average case. There are many results that improve the SEL algorithm and take advantage of the geometric properties of the dynamic programming array (i.e. values in neighbor cells differ at most by one) [91] in order to compute kn instead of mn entries. For example, Ukkonen [92] developed an algorithm which is called CUTOFF whose expected running time is $O(nk)$, by computing only a part of the dynamic programming array.

Subsequently new algorithms were developed that are based on diagonal transition approach. The basic idea of the diagonal transition algorithms is the fact that the diagonals of the dynamic programming array are monotonically increasing. The algorithm is based on computing in constant time the positions where the values along the diagonals are incremented. Therefore, there are four algorithms which are based on diagonal transition approach: Brute Force, Landau - Vishkin [50, 51], Galil - Park [27] and Ukkonen - Wood [93]. In our study we include the Galil - Park algorithm (in short, GP). The preprocessing phase of GP algorithm takes $O(m^2)$ space and time and the searching phase is $O(kn)$ time in the worst case or the average case. It uses reference triples that represent matching substrings of the pattern and the text as the Landau - Vishkin algorithm.

Finally, the Chang - Lampe algorithm [17] (in short, CL) is a variation of the dynamic programming and is also very efficient and practical algorithm. This adaptation of the simple dynamic programming approach is based on a "column partition" approach, and has expected time in $O(kn/\sqrt{|\Sigma|})$. The running time for preprocessing and the space requirement for this algorithm is $O(m|\Sigma|)$.

Deterministic Finite Automata Approach

Although this approach is rather old has received little attention. It is based on reexpressing the problem by mean of an automaton. The basic idea is to convert the general automaton into a deterministic one and reduce the states and the memory requirements.

Ukkonen devised an algorithm who proposed the idea of such a deterministic finite automaton (DFA) [92]. However, this algorithm has the disadvantage that a large number of automaton states may be generated. As a result, we have large time and space requirements which may limit the applicability of this algorithm.

Later, Wu et al. looked again into this problem [97]. The idea was to trade some time for space using a Four Russians technique [2] and give an $O(kn/logn)$ expected time algorithm which is a log factor improvement over the CUTOFF $O(kn)$ expected time algorithm, and an $O(mn/logn)$ time algorithm in the worst case using $O(n + m|\Sigma|/logn)$ space for the universal lookup array. The running time for preprocessing is $O(m|\Sigma|)$.

Finally, [47] and [68] proposed another way to reduce space requirements. It is an adaptation of [6], who first proposed it for the Hamming distance. The idea was to build the automaton in lazy form, i.e. build only the states and transitions actually reached in the processing of the text. This algorithm has running time $O(n + m\,min(r,n))$ where r is the total number of transitions in the complete automaton and it requires $O(min(|\Sigma|,n)min(m,|\Sigma|))$ space in the worst case. However, the average time complexity for this algorithm is $O(n + mr(1 - e^{-n/r}))$ where r is the total number of transitions in the complete automaton.

Filtering Approach

This method is much newer trend and it is currently very active. It is based on finding fast algorithms to discard large areas of the text that cannot match and apply another algorithm in the rest, using the simple dynamic programming approach.

First, Tarhio - Ukkonen [88] have devised an approximate string searching algorithm (in short, TUD) that tried to use Boyer - Moore - Horspool techniques [3, 37] to filter the text. The preprocessing phase has $O((k+|\Sigma|m)$ time and the space which is required by this algorithm is $O(|\Sigma|m)$. The searching phase of the TUD algorithm has $O(mn/k)$ time in the worst case and $O((|\Sigma|/|\Sigma| - 2k)kn(k/(|\Sigma| + 2k^2) + 1/m))$ in the average case.

Navarro [67] developed an algorithm (in short, COUNT), of [43] and [31] which is a filter based on counting matching positions. In other words, the key idea is to search for substrings of the text whose distribution of characters differs from the distribution of characters in the pattern at most as much as it is possible under k differences. The preprocessing phase of the COUNT algorithm has $O(|\Sigma|+m)$ time and the searching phase has $O(n)$ time if the number of verifications is negligible. Finally, this algorithm uses $O(|\Sigma|)$ space.

Wu and Manber [96] proposed a simple filter which is called pattern partition approach. This approach is based on the following fact: an occurrence with at most k differences of a pattern of length m implies that at least one substring of length r in the pattern matches a substring of the text occurrence exactly, where $r = \lceil m/(k+1) \rceil$. There are many ways to use this idea. Perhaps the simplest one, used in [96], is to search for the first $k+1$ consecutive blocks of size r of the pattern p. If any of the blocks is an exact match, we try to extend the match, checking if there are at most k differences. This idea was used in conjunction with the extension of the SO algorithm [5] to string matching with differences. The combination of the pattern partition approach with the SO algorithm we called MULTIWM. This algorithm has $O(mn/w)$ time complexity. Further, this algorithm is limited to $m \leq 31$.

Finally, Baeza et al. [11] suggested a algorithm (in short, BYPEP) which combines the pattern partition approach with traditional multiple string searching algorithms. The simplest algorithm is to build an Aho - Corasick machine [1] (the extension of the KMP algorithm [45, 61] to search for multiple patterns) for the $k+1$ blocks of length r. For every match found, we extend the match, checking if there are at most k differences, by using the standard dynamic programming algorithm to check the edit distance between two strings. This algorithm with the AC machine has $O(n)$ expected search time for $k \leq O(m/logm)$ using $O(m^2)$ extra space. Moreover, the above algorithm of the searching phase can be improved by using multiple string searching algorithm based on the Boyer - Moore algorithm [20].

Bit-Parallelism Approach

We have seen this technique is applied to k-mismatches problem. Therefore, this approach can be applied similar way to k-differences problem. There are two main alternatives: parallelization of the non-deterministic finite automaton (NFA) and parallelization of the dynamic programming array.

Wu and Manber algorithm [96] (in short, WM) uses this approach to simulate the automaton by rows. This algorithm has a preprocessing phase which requires $O(m|\Sigma| + k\lceil m/w \rceil)$ time. Then the searching phase runs in $O(kn\lceil m/w \rceil)$ time in the worst and average case which is $O(kn)$ for patterns typical in text searching (i.e. $m \leq w$). Moreover, this algorithm requires $O(m|\Sigma|)$ space. This algorithm is limited to $m \leq 31$.

Baeza et al. [8–10] proposed an another algorithm (in short, BYN) which parallelizes the NFA by diagonals using bits of the computer word. The preprocessing phase of the BYN algorithm takes $O(|\Sigma| + mmin(m, |\Sigma|))$ time and it requires $O(|\Sigma|)$ space. The search phase needs $O(n)$ time in the worst and average case. This algorithm is limited to $m \leq 9$ for $w = 32$ bits.

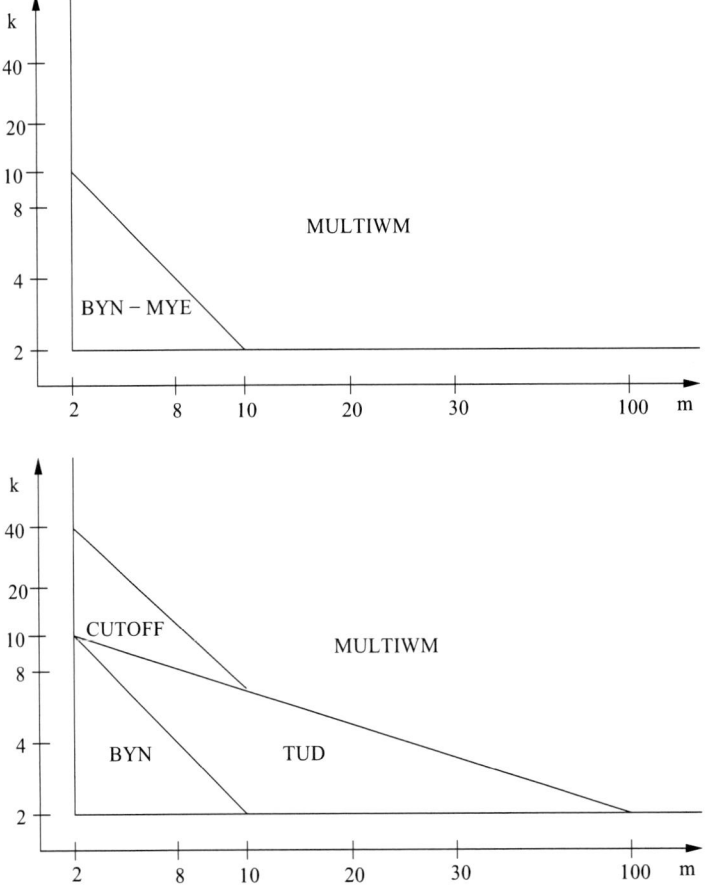

Figure 3.2. The areas where each string matching algorithm is best. English text is on top and binary text on the bottom.

Finally, Myers [65, 66] developed an algorithm (in short, MYE) which is based on bit parallel simulation of the dynamic programming array. The parallelization has optimal speedup, and the time complexity is $O(kn/w)$ on average and $O(mn/w)$ in the worst case. The preprocessing phase of the MYE algorithm requires $O(m|\Sigma|)$ time and $O(|\Sigma|)$ space. This algorithm is limited to $m \leq 31$.

3.2.3. Experimental Map

We now present a map of the most efficient string matching algorithms with k differences. To give an idea of the areas where each algorithm dominates, Figure 3.2 shows the cases of English text and binary text. We must noted that the bit-parallelism algorithms which are occurred in the map correspond to a word size of $w = 32$ bits.

Chapter 4

Parallel and Distributed Computing

Very often computational applications need more computing power than a sequential computer can provide. One way of overcoming this limitation is to improve the operating speed of processors and other components so that they can offer the power required by computationally intensive applications. Even though this is currently possible to certain extent, future improvements are constrained by the speed of light, thermodynamic laws, and the high financial costs for processor fabrication. A viable and cost-effective alternative solution is to connect multiple processors together and coordinate their computational efforts. The resulting systems are popularly known as parallel computers, and they allow the sharing of a computational task among multiple processors.

As Pfister [75] points out, there are three ways to improve performance:

- Work harder,
- Work smarter, and
- Get help.

In terms of computing technologies, the analogy to this mantra is that working harder is like using faster hardware (high performance processors or peripheral devices). Working smarter concerns doing things more efficiently and this revolves around the algorithms and techniques used to solve computational tasks. Finally, getting help refers to using multiple computers to solve a particular task.

4.1. Parallel Computer Architectures

The approximate string matching algorithms can be implemented to run efficiently on various types of hardware with the ability to perform several operations simultaneously. There is a wide range of different hardware available on which the algorithms can be implemented. The hardware can be divided into a group of general purpose parallel computers which can be used for many different kinds of computations and a group of special purpose parallel hardware specifically designed for performing single algorithms.

4.1.1. General Purpose Parallel Computers

General purpose computers with parallel processing capabilities usually contain a number of connected processors, ranging from dual-CPU workstations to clusters. Parallel computers are broadly categorized into shared memory and distributed memory computers based on how all the processors are coupled to the main memory.

In shared memory computers, all the processors are connected to a single global memory; all the processors have access to this global memory; also called as the tightly-coupled multiprocessor system. The communication between processors in this computer, takes place through the shared memory; modification of the data stored in global memory by one processor is visible to all other processors. Domimant representative shared memory systems are SGI, SMP (Symmetric Multi-Processing), dual-core etc.

In distributed memory computers, all the processors have their own local memory; also called as the loosely-coupled multicomputer system. The communication between processors in this computer, takes place through the interconnection network. The network connecting processors can be configured to tree, mesh, cube, torus, etc. Dominant representative distributed memory computers are IBM's SP/2, Intel's Paragon, etc.

Recently, a representative distributed memory system is the cluster of workstations connected by an Ethernet network is loosely connected workstations, is very interesting for string matching algorithms. In the next section, we will examine the cluster architecture in more detail.

4.1.2. Special Purpose Parallel Hardware

Special purpose hardware is usually built using either custom VLSI (Very Large Scale Integration) technology or FPGA (Field-Programmable Gate Arrays).

The VLSI technology have made possible the development of application specific arrays processor for complex and computationally intensive problems. The characteristics of parallelism, concurrency, pipelining, modularity and regularity have become standard in VLSI designs. The application specific arrays processors can provide the fastest means of running a particular algorithm with very high processing element (PE) density. However, they are limited to a single algorithm, and thus cannot supply the flexibility necessary to run the wide variety of different algorithms.

Reconfigurable systems are based on programmable logic such as FPGAs or custom-designed arrays. They are generally slower and have lower PE densities than application specific array processor architectures. They are flexible but the configuration must be changed for each algorithm, which is generally more complicated than writing new code for a programmable architecture. Solutions based on FPGAs have the additional advantage that they can be regularly upgraded to state-of-the-art technology. This makes FPGAs a very attractive alternative to array processor architectures.

4.2. Towards Low Cost Parallel Computing and Motivations

The use of parallel processing as a means of providing high performance computational facilities for large-scale and grand-challenge applications has been investigated widely. Until recently, however, the benefits of this research were confined to the individuals who had access to such systems. The trend in parallel computing is to move away from specialized traditional supercomputing platforms, such as the Cray/SGI T3E, to cheaper, general purpose systems consisting of loosely coupled components built up from single or multiprocessor PCs or workstations. This approach has a number of advantages, including being able to build a platform for a given budget which is suitable for a large class of applications and workloads.

The use of clusters to prototype, debug, and run parallel applications is becoming an increasingly popular alternative to using specialized, typically expensive, parallel computing platforms. An important factor that has made the usage of clusters a practical proposition is the standardization of many of the tools and utilities used by parallel applications. Examples of these standards are the message passing library MPI and data-parallel language HPF. In this context, standardization enables applications to be developed, tested, and even run on network of workstations (NOW), and then at a later stage to be ported, with little modification, onto dedicated parallel platforms where CPU-time is accounted and charged.

The following list highlights some of the reasons NOW is preferred over specialized parallel computers [15, 16]:

- Individual workstations are becoming increasingly powerful. That is, workstation performance has increased dramatically in the last few years and is doubling every 18 to 24 months. This is likely to continue for several years, with faster processors and more efficient multiprocessor machines coming into the market.

- The communications bandwidth between workstations is increasing and latency is decreasing as new networking technologies and protocols are implemented in a LAN.

- Workstation clusters are easier to integrate into existing networks than special parallel computers.

- Typical low user utilization of personal workstations.

- The development tools for workstations are more mature compared to the contrasting proprietary solutions for parallel computers, mainly due to the nonstandard nature of many parallel systems.

- Workstation clusters are a cheap and readily available alternative to specialized high performance computing platforms.

- Clusters can be easily grown; node's capability can be easily increased by adding memory or additional processors.

Clearly, the workstation environment is better suited to applications that are not communication-intensive since a LAN typically has high message start-up latencies and low bandwidths. If an application requires higher communication performance, the existing commonly deployed LAN architectures, such as Ethernet, are not capable of providing it.

4.3. Architecture of a Cluster Computer

A cluster is a type of parallel or distributed processing system, which consists of a collection of interconnected stand-alone computers working together as a single, integrated computing resource.

A computer node can be a single or multiprocessor system (PCs, workstations, or SMPs) with memory, I/O facilities, and an operating system. A cluster generally refers to two or more computers (nodes) connected together. The nodes can exist in a single cabinet or be physically separated and connected via a LAN. An interconnected (LAN-based) cluster of computers can appear as a single system to users and applications. Such a system can provide a cost-effective way to gain features and benefits (fast and reliable services) that have historically been found only on more expensive proprietary shared memory systems. The typical architecture of a cluster is shown in Figure 4.1.

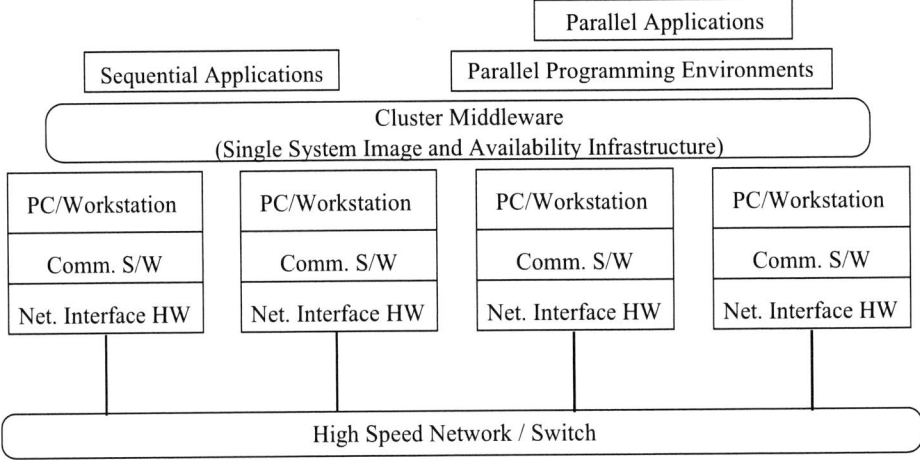

Figure 4.1. Architecture of a cluster of workstations [15].

The following are some prominent components of cluster computers [15]:

- Multiple High Performance Computers (PCs, Workstations, or SMPs)

- State-of-the-art Operating Systems (Layered or Micro-kernel based)

- High Performance Networks/Switches (such as Gigabit Ethernet and Myrinet)

- Network Interface Cards (NICs)

- Fast Communication Protocols and Services (such as Active and Fast Messages)

- Cluster Middleware (Single System Image (SSI) and System Availability Infrastructure)

    - Hardware (such as Digital (DEC) Memory Channel, hardware DSM, and SMP techniques)
    - Operating System Kernel or Gluing Layer (such as Solaris MC and GLUnix)
    - Applications and Subsystems
        * Applications (such as system management tools and electronic forms)
        * Runtime Systems (such as software DSM and parallel file system)
        * Resource Management and Scheduling software (such as LSF (Load Sharing Facility) and CODINE (COmputing in DIstributed Networked Environments))

- Parallel Programming Environments and Tools (such as compilers, PVM (Parallel Virtual Machine), and MPI (Message Passing Interface))

- Parallel or Distributed Applications

The network interface hardware acts as a communication processor and is responsible for transmitting and receiving packets of data between cluster nodes via a network/switch.

Communication software offers a means of fast and reliable data communication among cluster nodes and to the outside world. Often, clusters with a special network/switch like Myrinet use communication protocols such as active messages for fast communication among its nodes. They potentially bypass the operating system and thus remove the critical communication overheads providing direct user-level access to the network interface.

The cluster nodes can work collectively, as an integrated computing resource, or they can operate as individual computers. The cluster middleware is responsible for offering an illusion of a unified system image (single system image) and availability out of a collection on independent but interconnected computers.

Programming environments can offer portable, efficient, and easy-to-use tools for development of applications. They include message passing libraries, debuggers, and profilers. It should not be forgotten that clusters could be used for the execution of sequential or parallel applications.

4.4. Programming Environment and Tools

The availability of standard programming tools and utilities have made clusters a practical alternative as a parallel processing platform. In this section we discuss a few of the most popular tools.

4.4.1. Threads

Threads are a popular paradigm for concurrent programming on uniprocessor as well as multiprocessors machines. On multiprocessor systems, threads are primarily used to simultaneously utilize all the available processors. In uniprocessor systems, threads are used to utilize the system resources effectively. This is achieved by exploiting the asynchronous behavior of an application for overlapping computation and communication. Multithreaded applications offer quicker response to user input and run faster. Unlike forked process, thread creation is cheaper and easier to manage. Threads communicate using shared variables as they are created within their parent process address space.

Threads are potentially portable, as there exists an IEEE standard for POSIX threads interface, popularly called pthreads [72]. The POSIX standard multithreading interface is available on PCs, workstations, SMPs, and clusters. Threads have been extensively used in developing both application and system software.

4.4.2. Message Passing

Message passing libraries allow efficient parallel programs to be written for distributed memory systems. These libraries provide routines to initiate and configure the messaging environment as well as sending and receiving packets of data. Currently, the two most popular high-level message-passing systems for scientific and engineering application are the PVM (Parallel Virtual Machine) [28] from Oak Ridge National Laboratory, and MPI (Message Passing Interface) defined by MPI Forum [86].

MPI is the most popular message passing library that be used to develop portable message passing programs using either C or Fortran. The MPI standard defines both the syntax as well as the semantics of a core set of library functions that are very useful in writing message passing programs. MPI was developed by a group of researchers from academia and industry and has enjoyed wide support by almost all the hardware vendors. Vendor implementations of MPI are available on almost all parallel systems. The MPI library contains over 125 functions but the number of key concepts is much smaller. These functions provide support for starting and terminating the MPI library, getting information about the parallel computing environment, point-to-point and collective communications.

4.5. Parallel Programming Models

It has been widely recognized that parallel applications can be classified into some well defined programming models. A few programming models are used repeatedly to develop many parallel programs. Each model is a class of algorithms that have the same control structure. In the literature, there are many models that are used in parallel programming. However, for the further detail for parallel programming models, the reader is referred to book [16]. In this section, we present a brief overview of the two popular programming models, including master-worker and pipeline.

4.5.1. The Master-Worker Model

The master-worker model consists of two entities: master and multiple workers. The master is responsible for decomposing the problem into small tasks (and distributes these tasks among a farm of worker processes), as well as for gathering the partial results in order to produce the final result of the computation. The worker processes execute in a very simple cycle: get a message with the task, process the task, and send the result to the master. Usually, the communication takes place only between the master and the workers.

Master-worker may either use static load-balancing or dynamic load-balancing. In the first case, the distribution of tasks is all performed at the beginning of the computation, which allows the master to participate in the computation after each worker has been allocated a fraction of the work. The allocation of tasks can be done once or in a cyclic way. The other way is to use a dynamically load-balanced master/worker paradigm, which can be more suitable when the number of tasks exceeds the number of available processors, or when the number of tasks is unknown at the start of the application, or when the execution times are not predictable, or when we are dealing with unbalanced problems.

The master-worker model can be generalized to the hierarchical or multi-level master-worker model in which the top-level master feeds large chunks of tasks to second-level masters, who further subdivide the tasks among their own workers and may perform part of the work themselves. This model is generally equally suitable to shared memory or message-passing paradigms since the interaction is naturally two-way; i.e., the master knows that it needs to give out work and workers know that they need to get work from the master.

While using the master-worker model, care should be taken to ensure that the master does not become a bottleneck, which may happen if the tasks are too small (or the workers are relatively fast). The granularity of tasks should be chosen such that the cost of doing work dominates the cost of transferring work and the cost of synchronization. Asynchronous interaction may help overlap interaction and the computation associated with work generation by the master. It may also reduce waiting times if the nature of requests from workers is non-deterministic.

4.5.2. The Pipeline Model

In the pipeline model, a stream of data is passed on through a succession of processes, each of which perform some task on it. This simultaneous execution of different programs on a data stream is called stream parallelism. With the exception of the process initiating the pipeline, the arrival of new data triggers the execution of a new task by a process in the pipeline. The processes could form such pipelines in the shape of linear or multidimensional arrays, trees or general graphs with or without cycles. A pipeline is a chain of producers and consumers. Each process in the pipeline can be viewed as a consumer of a sequence of data items for the process preceding it in the pipeline and as a producer of data for the process following it in the pipeline. The pipeline does not need to be a linear chain; it can be a directed graph. The pipeline model usually involves a static mapping of tasks onto processes.

Load balancing is a function of task granularity. The larger the granularity, the longer it takes to fill up the pipeline, i.e. for the trigger produced by the first process in the chain to propagate to the last process, thereby keeping some of the processes waiting. However, too fine a granularity may increase interaction overheads because processes will need to interact to receive fresh data after smaller pieces of computation. The most common interaction reduction technique applicable to this model is overlapping interaction with computation.

4.6. Mapping Algorithms to Array Processors

The major emphasis of VLSI system design is to reduce the overall interconnection complexity and to keep the overall achitecture highly regular, parallel and pipelined. It stresses the importance of local communication in the array processor. Systolic and wavefront arrays are a class of pipelined array architectures. A systolic array is a network of processors which rhythmically compute and pass data through the system [48]. In constrast, a wavefront array is an asynchronous, selft-timed, data-driven computation array [48]. They are well suited to VLSI implementation, because of their properties of modularity, regularity, local interconnection and pipelining.

In the following, a systematic mapping methodology for deriving array processors is briefly reviewed. Our design begins with a data dependence graph to express the parallelism. Next, this graph will be mapped onto an array processor. Similar methods can be used to map algorithms onto systolic/wavefront arrays. For more details, the reader is referred to [48].

4.6.1. Deriving Dependence Graph from Given Algorithms

A data dependence graph is a directed graph which specifies the data dependencies of an algorithm. A data dependence graph consists of nodes and edges, where each node corresponds to an operation or computation and each edge corresponds to a data dependence between computations. For regular and recursive algorithms, the dependence graphs will also be regular and can be represented by a grid model. This model may contains the broadcast dependencies such as a node broadcasts a value to other nodes of the graph. Therefore, we should be transformed the original dependence graph to local dependence graph so that one contains local dependencies between computations. Design of a locally linked dependence graph is a critical step in the design of processor arrays.

4.6.2. Mapping Dependence Graph onto Array Processors

Two tasks are involved in mapping a dependence graph onto a array processors. The first task is processor assignment and the second is schedule assignment [48]. It is common to use a linear projection for processor assignment, in which nodes of the dependence graph along a straight line are projected (assigned) to a processor or cell in the processor array such as in Figure 4.2 left. A linear scheduling is a most popular schedule assignment, in which nodes on a hyperplane in the dependence graph are scheduled to be processed at the same time step such as in Figure 4.2 right.

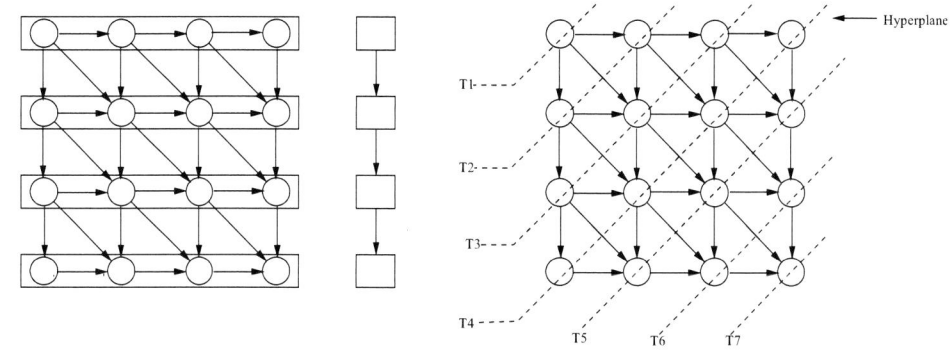

Figure 4.2. A linear projection (left) and a linear schedule and its hyperplanes (right).

Chapter 5

MPI Implementations of Exact and Approximate String Matching

A number of different designs for general-purpose parallel computers for performing exact/approximate string matching and similar problems have been proposed and implemented. In [24] a exact string matching implementation have been proposed and results are reported on a transputer based architecture. Further, in [44] the sequence comparison algorithms have been implemented to a variety of parallel computers, i.e. shared and distributed memory architectures. Finally, in [13, 14, 18, 41, 46, 52, 99] presented parallelisations of biological sequence alignment algorithms on a homogenous cluster of workstations. Some surveys on the fine-grain and coarse-grain parallelisation of the sequence analysis algorithms can be found in [78, 90, 98].

We follow master-worker programming model to develop our parallel and distributed approximate string matching implementations on a cluster of heterogeneous workstations using MPI library [30, 74, 86, 95]. This model consists of a master workstation and a collection of worker workstations. The master workstation is used to partition a given text collection into a set of several smaller subtext collections and distribute them to all worker workstations and to collect the local results from the worker workstations. The worker workstations are mainly performed a sequential exact and approximate string matching algorithm on their respective subtext collections [62, 63, 71]. Static and dynamic master-worker strategies are implemented and presented in next subsections.

5.1. The MPI Static Master-Worker Implementation

In order to present the static master-worker implementation we make the following assumptions: First, the workstations have an identifier myid and are numbered from 1 to p, second the documents of our text collection are distributed among the various workstations and stored on their local disks, and finally the pattern and the number of errors k are stored in main memory to all workstations. The partitioning strategy of this approach is to partition the entire text collection into a number of subtext collections according to the number of workstations allocated. The size of each subtext collection contains $\lceil n/p \rceil + m - 1$ successive characters of the complete text collection. There is an overlap of $m - 1$ pattern characters between successive subtexts, i.e. a reduance of $p(m-1)$ characters. Therefore, the static master-worker implementation that is called P1, is composed of four phases. In first phase, the master broadcasts the pattern string and the number of errors k to all the workers. In second phase, each worker reads its subtext collection from the local disk in main memory. In third phase, each worker performs character comparisons using a local sequential exact and approximate string matching algorithm to generate the number of occurrences. In fourth phase, the master collects the number of occurrences from each worker. This entire implementation is constructed so that alternative sequential exact and approximate string matching algorithms can be substituted quite easily [62, 63, 71].

The advantage of this simple approach is low communication overhead. This advantage was achieved, a priori, by the search computation, assigning each worker to search its own subtext independently without have to communicate with the other workers or the master. However, the main disadvantage is the possible load imbalance because of the poor partitioning technique. In the other words, there is a significant idle time for faster or more lightly loaded workstations in a heterogeneous environment.

5.2. The MPI Dynamic Master-Worker Implementation

In this subsection, we implement two versions of the dynamic master-worker model. The first version is based on the dynamic allocation of the subtexts and the second one is based on the dynamic allocation of the text pointers.

5.2.1. Dynamic Allocation of the Subtexts

The dynamic master-worker strategy that we adopted is a known parallelization strategy and is known as "workstation farm". Before, we present the dynamic implementation we make the following assumption: the entire text collection is stored on the local disk of the master workstation. The dynamic master-worker implementation that is called P2, is composed of six phases. In first phase, the master broadcasts the pattern string and the number of errors k to all workers. In second phase, the master reads from the local disk the several chunks of the text collection. The size of each chunk is $sb + m - 1$ successive characters where sb is the optimal block size. The block size is an important parameter which can be affect the overall performance. More specifically, this parameter is directly related to the I/O and communication factors. In third phase, the master sends the first chunks of the text collection to corresponding worker workstations. In fourth phase, each worker workstation performs a sequential exact and approximate string matching algorithm between the corresponding chunk of the text and the pattern in order to generate the number of occurrences. In fifth phase, each worker sends the number of occurrences back to master workstation. In sixth phase, if there are still any chunks of the text collection left, the master reads and distributes next chunks of the text collection to workers and loops back to fourth phase.

The advantage of this dynamic approach is low load imbalance, while the disadvantage is higher inter-workstation communication overhead.

5.2.2. Dynamic Allocation of Text Pointers

Before, we present the dynamic implementation with the text pointers we make the following assumptions: First, the complete text collection is stored on the local disks of all workstations and second, the master workstation has a text pointer that shows the current position in the text collection. The dynamic allocation of the text pointers that is called P3, is composed of six phases. In first phase, the master broadcasts the pattern string and the number of errors k to all workers. In second phase, the master sends the first text pointers to corresponding workers. In third phase, each worker reads from the local disk $sb + m - 1$ characters of the text starting from the pointer that receives. In fourth phase, each worker performs a sequential exact and approximate string matching procedure between the corresponding chunk of the text and the pattern in order to generate the number of occurrences. In fifth phase, each worker sends the result back to master. In sixth phase, if the text pointer does not reach the end of the text, then master updates the text pointers for the next position of next chunks of text and sends the pointers to workers and loops back to third phase.

The advantage of this simple implementation is that reduces the inter workstation communication overhead since each workstation in this scheme has an entire copy of the text collection on the local disk. However, this scheme requires more local space (or disk) requirements, but the size of the local disk in parallel and distributed architectures is large enough.

5.3. The MPI Hybrid Master-Worker Implmentation

Here, we develop a hybrid master-worker implementation that combines the advantages of three previous parallel implementations in order to reduce the load imbalance and communication overhead. This implementation is based on the optimal distribution strategy of the text collection which is performed statically. In the following subsection, we describe the optimal text distribution strategy and its implementation.

5.3.1. Text Distribution and Load Balancing

To avoid the slowest workstations to determine the parallel string matching time, the load should be distributed proportionally to the capacity of each workstation. The goal is to assign the same amount of time which may not correspond to the same amount of text collection.

To achieve a good balanced distribution among heterogeneous workstations, the amount of text distributed to each workstation should be proportional to its processing capacity compared to the entire network:

$$l_i = \frac{S_i}{\sum_{j=1}^{p} S_j} \qquad (5.1)$$

where S_i is the speed of the workstation i in the cluster. Therefore, the amount of the text collection that is distributed to each workstation M_i ($1 \leq i \leq p$) is $l_i * (n + m - 1)$ successive characters. There is an overlap of $m - 1$ pattern characters between successive subtexts.

The hybrid implementation that is called P4 is same as the P1 implementation but we use the optimal distribution method instead of the uniform distribution one.

Chapter 6

Description of Algorithms for Implementation

This section, we describe the computational intensive flexible approximate string matching algorithms for implementation onto array processors. This description will help to make parallelism and defining dependence graphs. In the following, two class of algorithms including dynamic programming and nondeterministic finite automata are briefly reviewed.

6.1. Dynamic Programming Algorithms

In this section, we cover the classical dynamic programming algorithms for the solution of the approximate string matching problem. In general, a string matching algorithm consists of two phases: the preprocessing and the searching phase. The preprocessing phase consists of gathering some information about the pattern, which can be used for fast implementation in the searching phase. On the other hand, the searching phase consists either of scanning the text or the construction of a dynamic programming matrix in order to find all exact or approximate occurrences of the pattern in the text. Therefore, this phase is based on the construction of a dynamic programming matrix D.

6.1.1. Preprocessing Phase

During the preprocessing phase the string p is encoded onto a bit-level memory map R of $(m \times |\Sigma|)$ bits, where $|\Sigma|$ is the size of the alphabet. The memory map can be seen as a two dimensional bit-level array where each row corresponds to a character of the pattern string and each column to a character of the alphabet. Therefore, column R_j^T, for $1 \leq j \leq |\Sigma|$, holds the information of the j-th character of the alphabet, which will be denoted as σ_j. The column R_j^T can be seen as a bit-level vector of m bits where the i-th bit, for $1 \leq i \leq m$, i.e. $R_{i,j}$ holds information concerning the j-th character σ_j and the i-th position of the search string. The basic information that can be recorded is whether the j-th character of the alphabet is the i-th character of the search string, that is whether $p_i = \sigma_j$. The preprocessing phase constructs the memory map R and the algorithm can be expressed in terms of the following piece of pseudocode.

for $j = 1$ to $|\Sigma|$ do
   for $i = 1$ to m do
      $R_{i,j} \leftarrow 1$
      if $p_i = \sigma_j$ then $R_{i,j} \leftarrow 0$

Taking as reference alphabet the UNICODE character set, it is straightforward to calculate the memory requirements of the bit map R. Thus $|\Sigma| = 64K$ and therefore the memory map of an m character search pattern requires $8m$ Kbytes. The overall space requirements are $\lceil m|\Sigma|/16 \rceil$ bytes, assuming that a character is encoded in two bytes. Further, the preprocessing phase can be performed in approximately $m|\Sigma|$ steps.

The algorithm performance can be improved if a specialized addressing is introduced such that a character is mapped directly to the appropriate column of the memory map. Such an addressing can take the form of a mapping function of a character ch of an alphabet Σ, $map(ch,\Sigma)$, returning the column number of the memory map. The additional details for this function are presented in [58].

6.1.2. Exact String Matching

The dynamic programming for the exact string matching is the same as the dynamic programming for the approximate string matching with k mismatches in which $k = 0$. See the section below.

6.1.3. Approximate String Matching with k Mismatches

We compute an $(m+1) \times (n+1)$ dynamic programming matrix $D_{0..m,0..n}$ such that $D_{i,j}$, $0 \leq i \leq m$ and $0 \leq j \leq n$ is the minimum Hamming distance between $p_{1..i}$ and a substring of t ending at t_j which has length i. So, the values of the matrix D are computed as follows:

for $i = 0$ to m do $D_{i,0} \leftarrow i$
for $j = 1$ to n do
   $D_{0,j} \leftarrow 0$
   $c \leftarrow map(t_j, \Sigma)$
   for $i = 1$ to m do
      $D_{i,j} \leftarrow D_{i-1,j-1} + R_{i,c}$

The expression $D_{i-1,j-1} + R_{i,c}$ corresponds to substituting t_j for p_i. Whenever $D_{m,j} \leq k$, $1 \leq j \leq n$ it means that there is approximate occurrence of pattern ending at position j in the text with number of mismatches less than or equal to k. This algorithm can be performed in $O(mn)$ steps and the space requirements are $\lceil mn/16 \rceil$ bytes. Another important observation is that the computation of column j depends only on the values of column $j-1$. Therefore, there is no need to store the whole matrix and space requirements are reduced to $\lceil m/16 \rceil$ bytes.

6.1.4. Approximate String Matching with k Differences

We compute an $(m+1) \times (n+1)$ dynamic programming matrix $D_{0..m, 0..n}$ such that $D_{i,j}$, $0 \leq i \leq m$ and $0 \leq j \leq n$ is the minimum edit distance (i.e. the minimum number of differences) between $p_{1..i}$ and any substring of t ending at t_j. The values of the matrix D are computed as follows [81, 92, 94]:

for $i = 0$ to m do $D_{i,0} \leftarrow i$
for $j = 1$ to n do
 $D_{0,j} \leftarrow 0$
 $c \leftarrow map(t_j, \Sigma)$
 for $i = 1$ to m do
  $D_{i,j} \leftarrow min(D_{i-1,j-1} + R_{i,c}, D_{i,j-1} + 1, D_{i-1,j} + 1)$

The three expressions in the *min* correspond respectively to substituting t_j for p_i, inserting t_j into the pattern, and deleting p_i from the pattern. Whenever $D_{m,j} \leq k$, $1 \leq j \leq n$ it means that there is an approximate occurrence of pattern ending at position j in the text with edit distance less than or equal to k differences. Further, this search computation is performed in $O(mn)$ steps and it requires $\lceil mn/16 \rceil$ bytes of space.

6.1.5. Approximate String Matching with k Differences Based on Myers Algorithm

We must mention that the dynamic programming matrix D has an important geometric property [59, 92]: adjacent entries along horizontal and vertical directions differ by 1, 0 or -1. Formally, we define the horizontal delta $\Delta h_{i,j}$ at (i,j) as $D_{i,j} - D_{i,j-1}$ and the vertical delta $\Delta v_{i,j}$ as $D_{i,j} - D_{i-1,j}$ for $1 \leq i \leq m$ and $1 \leq j \leq n$. We have $-1 \leq \Delta h_{i,j}, \Delta v_{i,j} \leq 1$ for $1 \leq i \leq m$ and $1 \leq j \leq n$.

We can thus replace the problem of computing D with the problem of computing the dynamic programming matrices Δh and Δv. The next task is to develop an understanding of how to compute the deltas in one column from those in the previous column. To start with, let us consider an individual cell of the dynamic programming matrix consisting of the square $(i-1, j-1), (i-1, j), (i, j-1)$ and (i, j). There are two horizontal and two vertical deltas - $\Delta v_{i,j}, \Delta v_{i,j-1}, \Delta h_{i,j}$ and $\Delta h_{i-1,j}$ - associated with the sides of this cell. Using the definition of the deltas and the basic recurrence for D-values we arrive at the following equation for $\Delta v_{i,j}$ in terms of $R_{i,c}, \Delta v_{i,j-1}$ and $\Delta h_{i-1,j}$ [66, 97]:
$\Delta v_{i,j} = min\{1, R_{i,c} - \Delta h_{i-1,j}, \Delta v_{i,j-1} - \Delta h_{i-1,j} + 1\}$.
Similarly for $\Delta h_{i,j}$:
$\Delta h_{i,j} = min\{1, R_{i,c} - \Delta v_{i,j-1}, \Delta h_{i-1,j} - \Delta v_{i,j-1} + 1\}$.
Therefore, the values of the matrices Δv and Δh are computed by the following piece of pseudocode [66, 97].

for $i = 1$ to m do $\Delta v_{i,0} \leftarrow 1$
$e_0 \leftarrow m$
for $j = 1$ to n do
    $\Delta h_{0,j} \leftarrow 0$
    $c \leftarrow map(t_j, \Sigma)$
    for $i = 1$ to m do
        $\Delta v_{i,j} \leftarrow min\{1, R_{i,c} - \Delta h_{i-1,j}, \Delta v_{i,j-1} - \Delta h_{i-1,j} + 1\}$
        $\Delta h_{i,j} \leftarrow min\{1, R_{i,c} - \Delta v_{i,j-1}, \Delta h_{i-1,j} - \Delta v_{i,j-1} + 1\}$
    $e_j \leftarrow e_{j-1} + \Delta h_{m,j}$

In order to calculate the difference of each column of the matrix Δv we maintain the value of $e_j = D_{m,j}$ as one computes the Δv_j's using the fact that $e_0 = m$ and $e_j = e_{j-1} + \Delta h_{m,j}$. If $e_j \leq k$ then there is approximate occurrence of pattern ending at position j in the text with edit distance less than or equal to k. Finally, the search time of this computation is $O(mn)$ steps.

6.2. Nondeterministic Finite Automata (NFA) Algorithms

In this section, we present the string matching algorithms which are based on the simulation of a nondeterministic finite automaton built from the pattern and using the text as input.

6.2.1. Algorithm Based on the Rows of the Automaton

This category of algorithms consists of two phases: the preprocessing and searching phase. The preprocessing phase is the same as the dynamic programming algorithm and the details are described in Section 6.1.1.. The searching phase is based on the simulation of a nondeterministic finite automaton by rows.

Exact String Matching

Consider the NFA for pattern of length $m = 4$ without errors, shown in Figure 6.1. Every column represents matching the pattern up to a given position. The self-loop at the initial state allows us to consider any character as a potential starting point of a match. This NFA has $m + 1$ states. We assign number i to the state in column i, $0 \leq i \leq m$. Initially, the active state is the column 0.

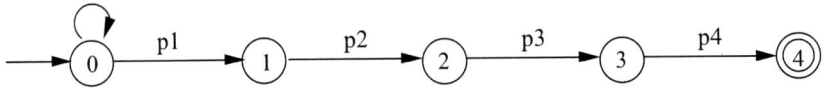

Figure 6.1. NFA for exact string matching.

We have a vector F^0 corresponding to the NFA automaton. F_i^0 is 0 if state i is active and 1 otherwise. The vector changes as each character of the text is read. The new values of vector $F_i'^0$, $0 \leq i \leq m$ after we read a new text character t_j, $1 \leq j \leq n$, are computed as follows [5]:

for $i = 0$ to m do
  $F_i^0 \leftarrow 1$
for $i = 0$ to 0 do
  $F_i^0 \leftarrow 0$
for $j = 1$ to n do
  $c \leftarrow map(t_j, \Sigma)$
  for $i = 1$ to m do
    $F_i'^0 \leftarrow F_{i-1}^0$ OR $R_{i,c}$
  for $i = 0$ to m do
    $F_i^0 \leftarrow F_i'^0$

The first term of formula F'^0 represents matching. If $F_m'^0 = 0$ then there is exact occurrence of a pattern in the text. Finally, this NFA simulation requires $O(mn)$ steps.

Approximate String Matching with k Mismatches

Consider the NFA for a pattern of length $m = 4$ with $k = 2$ mismatches, shown in Figure 6.2. Every row denotes the number of mismatches seen. The first one is 0, the second one is 1, an so on. Horizontal arrows represent matching a character, and solid diagonal arrows represent replacing the character of the text with the corresponding character in the pattern (that is, a mismatch). This NFA has $(m+1) \times (k+1)$ states. We assign number (l, i) to the state at row l and column i, $0 \leq l \leq k$, $0 \leq i \leq m$. Initially, the active states at row l are in columns from 0 to l.

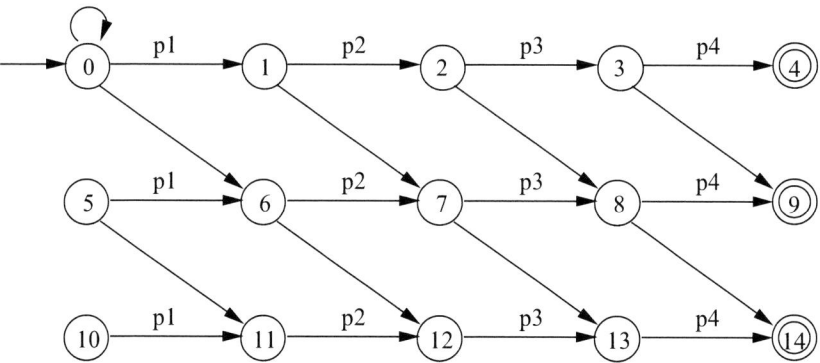

Figure 6.2. NFA for string matching with mismatches.

Consider the $k+1$ vectors F^l, $0 \leq l \leq k$ that corresponding the rows of the automaton. F_i^l is 0 if state (l, i) is active and 1 otherwise. The new values of vectors $F_i'^l$, $0 \leq l \leq k$, $0 \leq i \leq m$, after we read a new text character t_j, $1 \leq j \leq n$ are computed as follows:

for $l = 0$ to k do
  for $i = 0$ to m do
    $F_i^l \leftarrow 1$
for $l = 0$ to k do

```
for i = 0 to l do
    F_i^l ← 0
for j = 1 to n do
    c ← map(t_j, Σ)
    for i = 1 to m do
        F_i'^0 ← F_{i-1}^0 OR R_{i,c}
    for l = 1 to k do
        for i = 1 to m do
            F_i'^l ← (F_{i-1}^l OR R_{i,c}) AND F_{i-1}^{l-1}
    for l = 0 to k do
        for i = 0 to m do
            F_i^l ← F_i'^l
```

In Formula $F_i'^l$, the first term represents matching and the second term represents edit operation substitution. If $F_m'^k = 0$, then there is approximate occurrence of a pattern in the text with at most k mismatches. The above NFA simulation can be performed in $O(nmk)$ steps.

Approximate String Matching with k Differences

Consider now the NFA for a searching pattern of length $m = 4$ with $k = 2$ differences, shown in Figure 6.3. As for the case of only mismatches, every row denotes the number of differences seen. The structure of the automaton is similar to the case of mismatches adding two additional transitions per state. Dashed diagonal arrows represent deleting a character of the pattern (they are empty transition), while solid vertical arrows represent inserting a character into the pattern. Let F_i^l be 0 if the state on row l and column i is active. The new values of vectors $F_i'^l$, $0 \leq l \leq k$, $0 \leq i \leq m$, after we read a new text character t_j, $1 \leq j \leq n$, are computed as follows [96]:

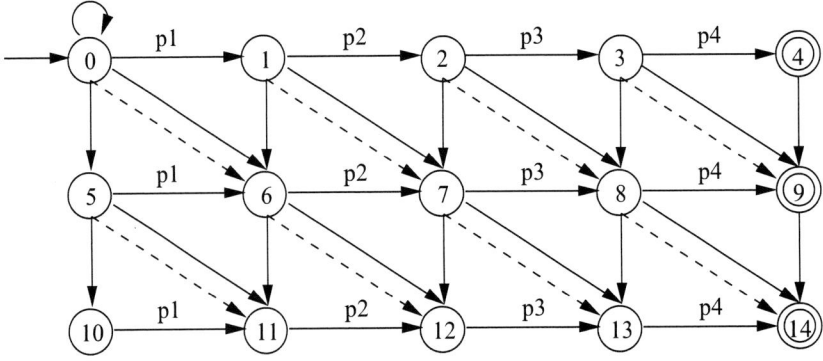

Figure 6.3. NFA for string matching with differences.

```
for l = 0 to k do
    for i = 0 to m do
```

$$F_i^l \leftarrow 1$$
for $l = 0$ to k do
   for $i = 0$ to l do
      $F_i^l \leftarrow 0$
for $j = 1$ to n do
   $c \leftarrow map(t_j, \Sigma)$
   for $i = 1$ to m do
      $F_i'^0 \leftarrow F_{i-1}^0$ OR $R_{i,c}$
   for $l = 1$ to k do
      for $i = 1$ to m do
         $F_i'^l \leftarrow (F_{i-1}^l$ OR $R_{i,c})$ AND F_i^{l-1} AND F_{i-1}^{l-1} AND $F_{i-1}'^{l-1}$
   for $l = 0$ to k do
      for $i = 0$ to m do
         $F_i^l \leftarrow F_i'^l$

In Formula F'', the first term represents matching, the second represents edit operation insert, the third represents edit operation substitution and the last represents edit operation delete. If $F_m'^k = 0$, then there is approximate occurrence of a pattern in the text with at most k differences. The above simulation requires $O(mnk)$ steps.

6.2.2. Algorithm based on the Columns of the Automaton

The searching phase of this algorithm is based on the simulation of the automaton by columns.

Exact String Matching

The NFA simulation by columns for exact string matching is the same as the NFA simulation for approximate string matching with k mismatches in which $k = 0$. See the section below.

Approximate String Matching with k Mismatches

Let us define the $m + 1$ numbers C_i, $0 \leq i \leq m$ on the range $0...k + 1$ that corresponds the columns of the automaton. Each value of C_i represents the smallest active state level per column. The new values of each column C_i' after we read a new text character are computed by the following piece of pseudocode [7]:

for $i = 0$ to m do
   $C_i \leftarrow i$
for $j = 1$ to n do
   $c \leftarrow map(t_j, \Sigma)$
   for $i = 1$ to m do
      $C_i' \leftarrow C_{i-1} + R_{i,c}$
   for $i = 0$ to m do

$$C_i \leftarrow C_i'$$

In Formula C_i', the first term represents either a match or a substitution. If $C_m' \leq k$, then there is approximate occurrence of pattern in the text with at most k mismatches. This solution requires $O(mn)$ steps. We recognize this solution as a variation of the well known dynamic programming approach to approximate string matching with k mismatches problem. At a given text character, if we collect the smallest active rows in each column we obtain the vertical vector of the dynamic programming matrix.

Approximate String Matching with k Differences

Similarly, we define the $m+1$ numbers C_i, $0 \leq i \leq m$ on the range $0...k+1$ that correspond to the columns of the automaton. Each value of C_i represents the minimum row of the NFA per column. The new values of each column C_i' after we read a new text character are computed by the following piece of pseudocode [7]:

```
for i = 0 to m do
    C_i ← i
    C_i' ← C_i
for j = 1 to n do
    c ← map(t_j, Σ)
    for i = 1 to m do
        C_i' ← min{C_{i-1} + R_{i,c}, C_i + 1, C_{i-1}' + 1}
    for i = 0 to m do
        C_i ← C_i'
```

In Formula C_i', the first term represents either a match or substitution, the second represents an insertion and the last represents a deletion. If $C_m' \leq k$, then there is approximate occurrence of pattern in the text with at most k differences. This search computation requires $O(mn)$ steps.

6.3. Extensions

In this section, we discuss some extensions that can support the previous string search algorithms.

6.3.1. Limited Expressions

A limited expression in [97] is a pattern that matches not only a single character but an arbitrary set of characters and it is a subset of the wealth of alternatives for extended patterns presented in [5, 96]. For convenience of notation, we use the symbols '*', '^', '[', and ']' to denote common types of symbols as follows (this is consistent with *grep* and *agrep*):

- A '*' don't care symbol can represent matching with any single character.

Description of Algorithms for Implementation

- A '^' complement symbol can represent matching with all characters except the one that is complemented.

- A pair of '[' and ']' defines a class symbol that allows matching with a subrange of characters.

To search for these limited expressions patterns, we need only to modify the preprocessing phase without additional search complexity. We have:

$$R_{i,j} = \begin{cases} 0, & if \quad p_i = *, \text{ for } 1 \leq j \leq |\Sigma| \\ 0, & if \quad p_i = \char`\^\sigma_1, \text{ for } 1 \leq j \leq |\Sigma| \text{ and } j \neq \text{map}(\sigma_1, \Sigma) \text{ for } 1 \leq i \leq m \\ 0, & if \quad p_i = [\sigma_1 \sigma_2], \text{ for } \text{map}(\sigma_1, \Sigma) \leq j \leq \text{map}(\sigma_2, \Sigma) \end{cases}$$

where σ_j is the j-th character of the alphabet.

Chapter 7

Data Dependence Graphs for Approximate String Matching Algorithms

Some surveys on arrays processors and architectures for exact/approximate string matching and related problems can be found in [19, 39, 40, 77]. A number of arrays processors have been proposed by several researchers for approximate string matching [19, 25, 33, 35, 38, 53, 54, 56–58, 60, 79, 80, 84].

In this section, we derive the data-dependence graphs for the searching phase of the dynamic programming and the NFA algorithms which are presented in Sections 6.1. and 6.2..

7.1. Dependence Graphs for the Dynamic Programming Algorithms

Figures 7.1, 7.2 and 7.3 show the dependence graphs and the parallel timing diagrams for the three dynamic programming algorithms. All nodes of the graphs 7.1, 7.2 and 7.3 which lie in the same row use copies of one and the same character of the input pattern string p. Similarly, all nodes which lie in one column use copies of one and the same character of text string t. For example, all nodes of the first row of the graphs are stored the first character p_1 of pattern p. Similarly, all nodes of the first column of the graphs are loaded the first character t_1 of the input text string t. Further, node (i, j) of graphs 7.1, 7.2 and 7.3 ($1 \leq i \leq m, 1 \leq j \leq n$) is assigned an entire row i of the bit-level memory map R for the character p_i of the pattern. In other words, the nodes of the first row of the graphs are assigned the first row of the R, corresponding to the first character of the pattern. The nodes of the second row of the graphs are assigned the second row of the R, corresponding to the second character of the pattern, and so on.

In Figure 7.1a, to compute an element of the dynamic programming matrix $D_{i,j}$ ($1 \leq i \leq m, 1 \leq j \leq n$), we need the previous diagonal calculated element $D_{i-1,j-1}$ as indicated in Figure 7.1b. Let each processor be responsible to compute one row of the dynamic programming matrix D. The dependence graph allows us to compute the elements on the same diagonal (from left-bottom to right-top) in parallel. This is shown by the dotted diag-

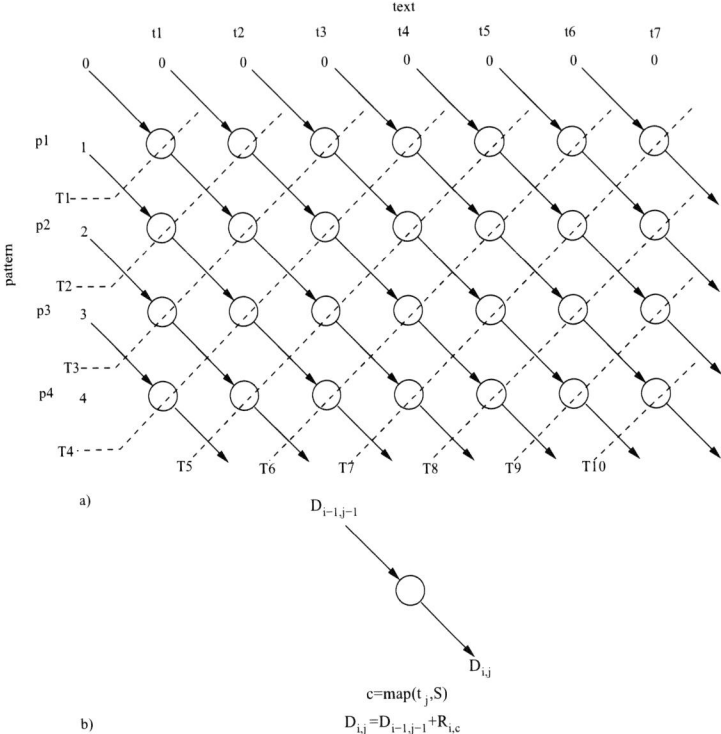

Figure 7.1. (a) Dependence graph and parallel timing diagram for string matching with mismatches (b) Computation.

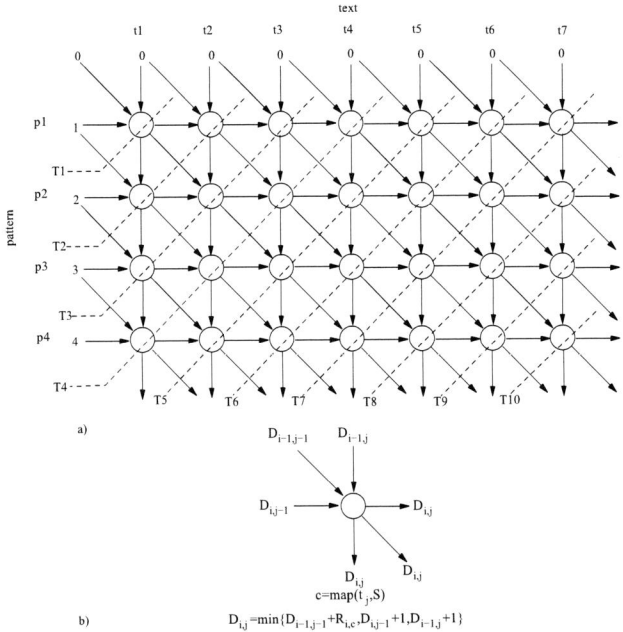

Figure 7.2. (a) Dependence graph and parallel timing diagram for string matching with differences (b) Min computation.

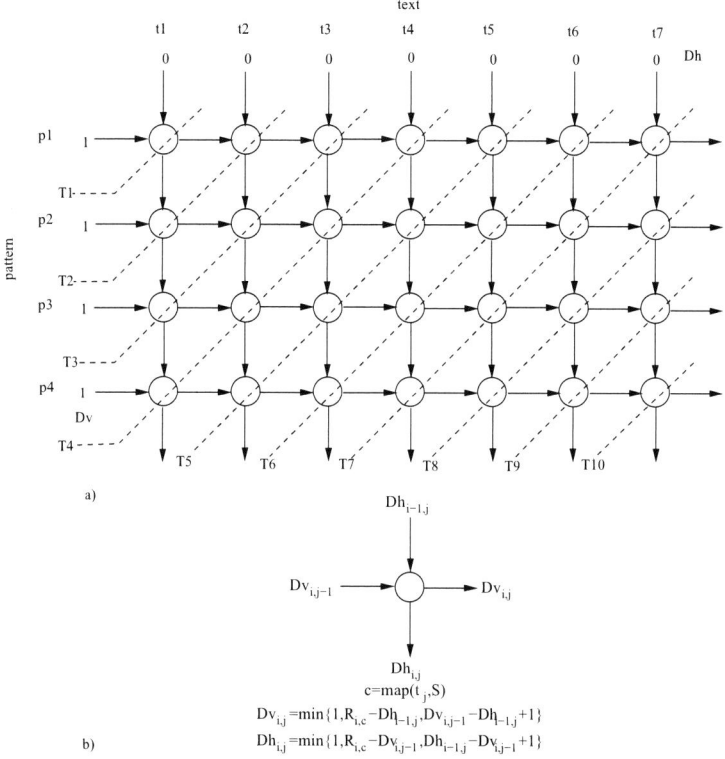

Figure 7.3. (a) Dependence graph and parallel timing diagram for string matching with differences is based on Myers algorithm (b) Min computations.

onal lines of Figure 7.1a. For instance, at time step T4 it computes the elements $D_{4,1}, D_{3,2}, D_{2,3}$, and $D_{1,4}$ concurrently.

In Figure 7.2a, to compute an element of the dynamic programming matrix $D_{i,j}$ ($1 \leq i \leq m, 1 \leq j \leq n$) we need three previously calculated values $D_{i-1,j}$, $D_{i,j-1}$ and $D_{i-1,j-1}$ as indicated in Figure 7.2b. Similar to the graph of Figure 7.1a, the dependence graph of Figure 7.2a allows us to compute all m elements on the same diagonal at the same time. This is shown by the dotted lines of Figure 7.2a.

Each node of Figure 7.3a computes the horizontal difference $\Delta h_{i,j}$ and the vertical difference $\Delta v_{i,j}$ ($1 \leq i \leq m, 1 \leq j \leq n$) of the dynamic programming matrices Δh and Δv respectively. In order to compute these differences, each node need to receive the previously calculated horizontal and vertical differences, $\Delta h_{i-1,j}$ and $\Delta v_{i,j-1}$ respectively, as indicated in Figure 7.3b. Finally, the computations along 45 degree diagonals are performed concurrently.

7.2. Dependence Graphs for the NFA Algorithms

Figures 7.4, 7.5 and 7.6 show the dependence graphs and the parallel timing diagrams for the string matching algorithms that simulate the NFA by rows. Node (i, j) of graphs 7.4, 7.5 and 7.6 ($1 \leq i \leq m, 1 \leq j \leq n$) is stored the character p_i of the pattern string p and character

t_j of input text string t. Moreover, each node (i,j) of the graphs is assigned a row i of the R for the pattern character p_i.

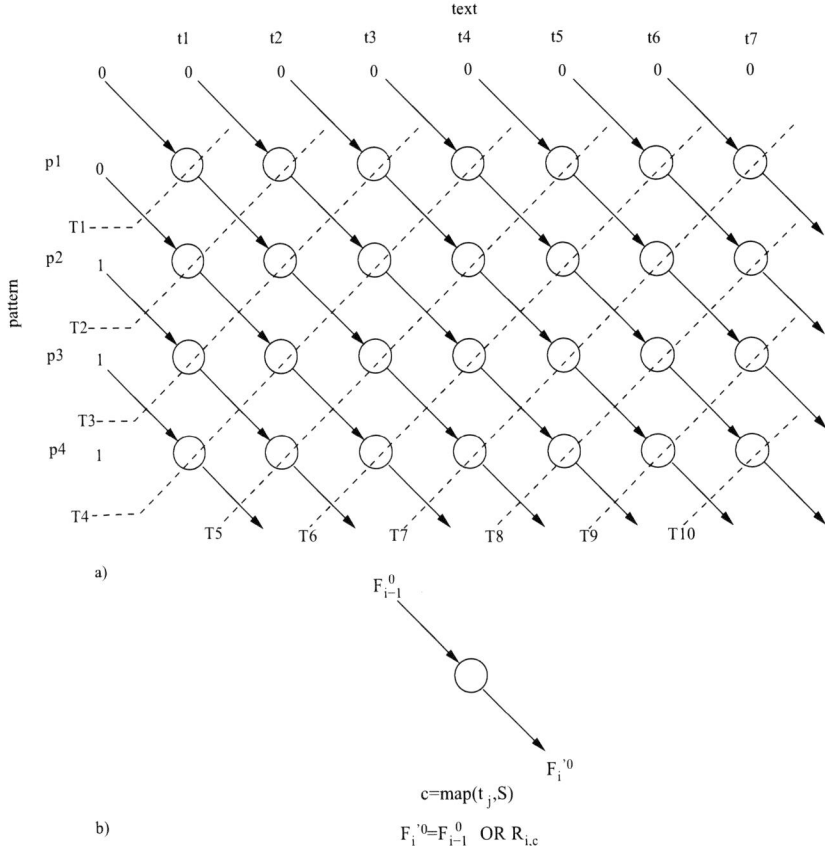

Figure 7.4. (a) Dependence graph and parallel timing diagram for exact string matching (b) Computation.

In Figure 7.4a, to calculate an element of vector $F_i'^0$ ($1 \leq i \leq m$) after we read a new text character t_j ($1 \leq j \leq n$), we need the previous value F_{i-1}^0 as indicated in Figure 7.4b. We suppose each processor is responsible to compute all nodes of each row of Figure 7.4a. We conclude from the dependence graph that all nodes which lie on the same diagonal (from left-bottom to right-top) can be performed at the same time as is shown in Figure 7.4a by dotted diagonal lines. Finally, we observe that the timing diagram of this graph is similar to the timing diagrams of Figures 7.1, 7.2 and 7.3.

Figure 7.5a shows the dependence graph for two levels, i.e. it allows string matching with 1 mismatches. Level $l=0$ corresponds to the calculations of vector $F_i'^0$, while the level $l=1$ corresponds to the calculations of vector $F_i'^1$. The computations and the parallel timing diagram for level $l=0$ are similar to Figure 7.4. Now, the level $l=1$ of Figure 7.5a, to calculate an element of the vector $F_i'^1$ ($1 \leq i \leq m$) after the reading the text character t_j ($1 \leq j \leq n$), we should receive two previously calculated values F_{i-1}^l and F_{i-1}^{l-1} as indicated in Figure 7.5b. Figure 7.5a shows the parallel timing diagram that the computations along 45 degree diagonals are performed concurrently. Finally, we can see from the timing diagrams

Data Dependence Graphs for Approximate String Matching Algorithms 43

for two levels that the nodes which lie on the same dotted diagonal lines can be performed in parallel. For example, time steps $T4(0)$ and $T4(1)$ for levels 0 and 1 respectively can be performed concurrently.

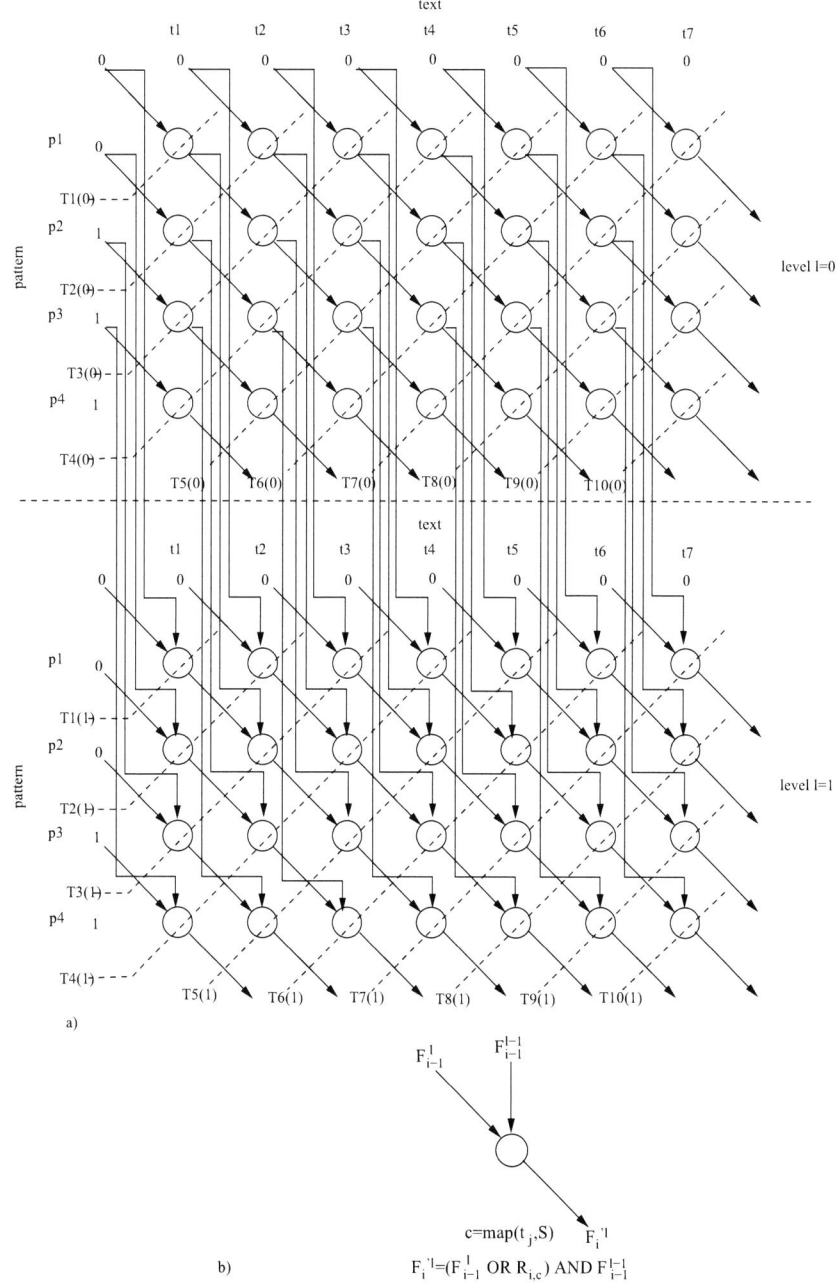

Figure 7.5. (a) Dependence graph and parallel timing diagram for string matching with mismatches (b) Computation.

Figure 7.6a shows the dependence graph for two levels, i.e., allow string matching with 1 difference. The computations and the parallel timing diagram for level $l = 0$ are similar to Figure 7.4. Each node of Figure 7.6a for level $l = 1$ computes the element of the vector F_i^{l1} ($1 \leq i \leq m$) after we read a new text character t_j ($1 \leq j \leq n$). In order to calculate this element, each node should receive four previously calculated elements $F_{i-1}^l, F_i^{l-1}, F_{i-1}^{l-1}$, and $F_{i-1}^{\prime l-1}$, as indicated in Figure 7.6b. From the dependence graph it is obtained that the nodes along 45 degree diagonals are performed parallel. Finally, from the dependence graph of Figure 7.6a we observe that the time steps $T4(0)$ and $T4(1)$ for example can be computed in parallel.

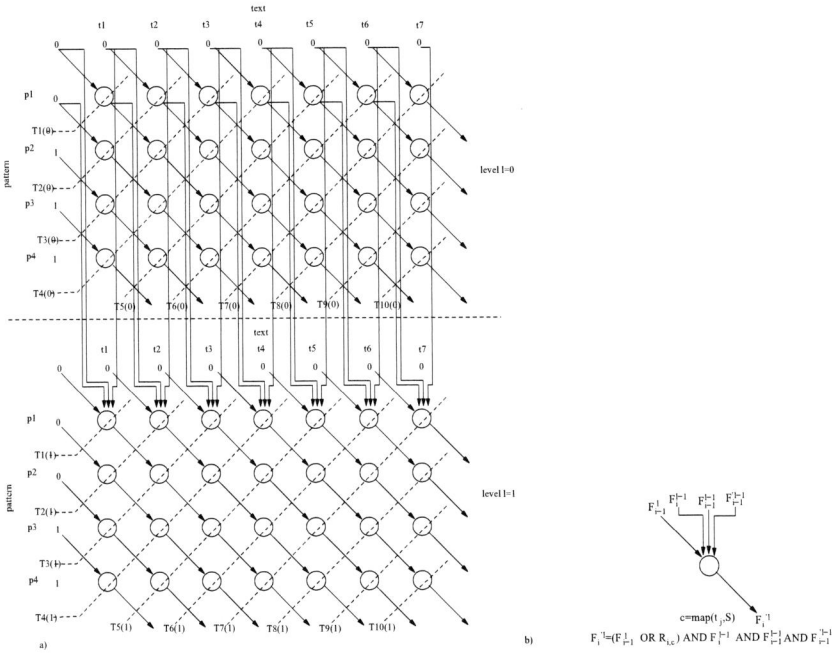

Figure 7.6. (a) Dependence graph and parallel timing for string matching with differences (b) Computation.

We do not derive the dependence graphs for the algorithms that simulate the NFA by columns because they are similar to the graphs of the dynamic programming algorithms.

Chapter 8

Mapping Approximate String Matching Algorithms onto Processor Arrays

Initially, the implementations of the searching phase of dynamic programming and the NFA algorithms are discussed on the processor array, which imposes the main computation load since $m << n$.

8.1. Processor Arrays for the Dynamic Programming Algorithms

The processor array architectures for original dependence graphs of Figures 7.1a, 7.2a and 7.3a can now be derived by specifying the processor and schedule assignments. The area complexity of the processor array algorithms is defined as the number of elementary cells (or PEs) being active at any time step. This information is given by the number of rows of the dependence graph whereas the time steps required are given by the number of diagonals of the graph.

It should be clear that the processor arrays are derived from projecting the horizontal axis of the dependence graph of Figures 7.1a, 7.2a and 7.3a and are shown in Figures 8.1, 8.2 and 8.3 for a generic problem with $m = 4$ and for arbitrary n and $|\Sigma|$.

Figure 8.1 shows a linear processor array consisting of m cells connected to each other via two communication channels, one transferring the binary representation of text characters and another transferring the byte-level delta results. Each row R_i, $1 \leq i \leq m$ of the bit-level memory map R is allocated to a PE, so that reading through the specialized addressing function $map(ch, \Sigma)$ produces a single bit per PE. This operation is realized in time $2m$. Each cell performs a full step of the computation as shown in Figure 8.1 on the right, i.e. both mapping and addition operation. A single time step is concluded with communication between adjacent PEs so that the partial results are pipelined towards the last cell. It is noted that the results travels along the same direction as the text characters but at half speed, that is with an intermediate delay between cells.

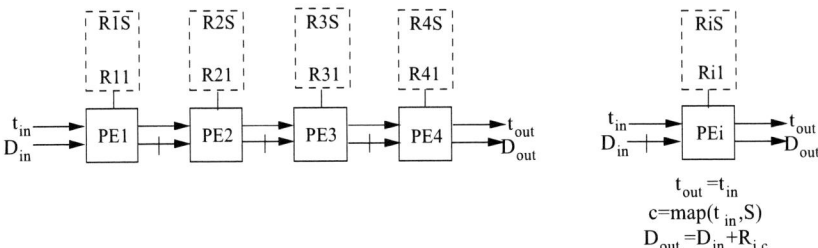

Figure 8.1. Processor array for string matching with mismatches.

Figure 8.2 shows a linear array of m cells connected to each other via three communication channels, the one transferring the binary representation of text characters, and the other two transferring the byte-level results. Each cell of the array is allocated as in the previous array a row of the R and register D which store the value $D_{i,j-1}$. Moreover, each cell performs the mapping, min and assignment operations as shown in Figure 8.2 right. Since each cell updates the D value, the latter is stored in register D and it is sent towards the next cell through the two results channels. It is noted that the results of the third communication channel flow from left to right in the same manner as the first two channels, but at half speed, that is with intermediate delay between cells.

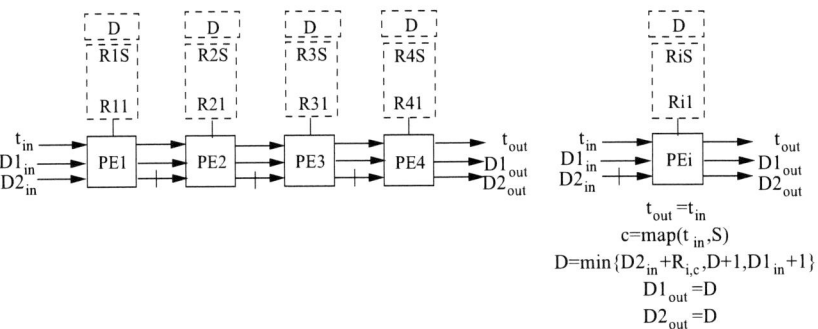

Figure 8.2. Processor array for string matching with differences.

Figure 8.3 shows a linear array connected to each other via two communication channels, one transferring the binary representation of text characters and another transferring the bit-level horizontal delta results. Each cell of the array is allocated a row of the R and two registers, Δv and aux which respectively store the current vertical difference $\Delta v_{i,j}$ and the previous difference $\Delta v_{i,j-1}$ after the processing of t_j. Also, each cell performs a full step of the computation as shown in Figure 8.3 on the right right, i.e. mapping, min and assignment operations. Therefore, each cell computes two difference values. One vertical difference is updated and stored in register Δv, while the other horizontal difference also is updated and is sent to the next cell. The text characters flow from left to right in the same manner as the partial results, without intermediate delay between cells compared to previous processor array designs.

The overall computation time for the three linear approaches is performed in $n+m-1$ time steps. Given the fact that usually $m \ll n$ it can be argued that the computation time is approximately n steps. The area required is m PEs for the three processor designs. Taking the example of the UNICODE character set the local memory requirements are 8Kbytes per cell.

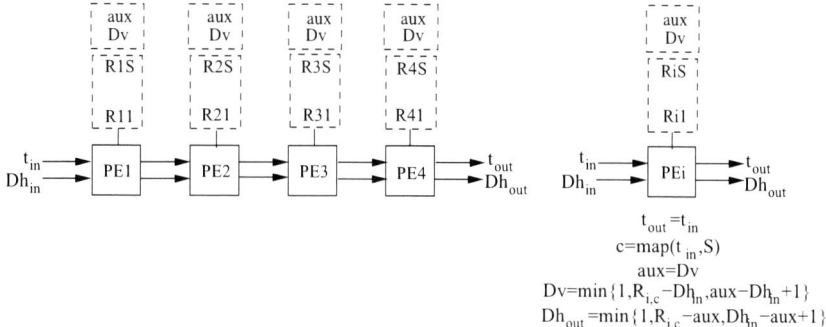

Figure 8.3. Processor array for string matching with differences is based on Myers algorithm.

8.2. Processor Arrays for the NFA Algorithms

Similarly, the processor arrays are derived from projecting the horizontal axis of the graph of Figures 7.4a, 7.5a and 7.6a and are shown in Figures 8.4, 8.5 and 8.6 for a generic problem with $m = 4$, $k = 1$ and for arbitrary n and $|\Sigma|$.

Figure 8.4 shows a linear processor array which is similar to the processor array design of Figure 8.1. However, the second communication channel transfers the bit-level results instead of the byte-level results. Further, each cell performs the logical operation instead of the addition calculation as shown in Figure 8.4 right, i.e. both mapping and OR operations. Therefore, each cell updates the result, which is sent to the next cell.

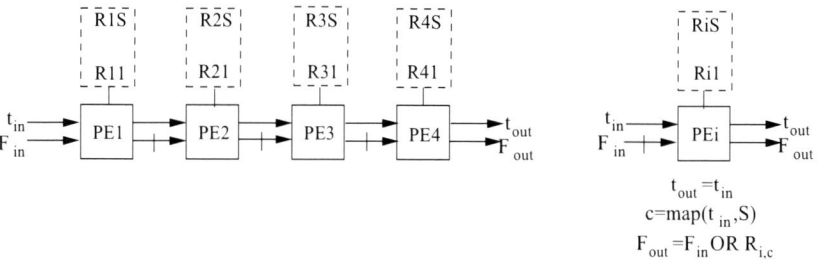

Figure 8.4. Processor array for exact string matching.

Figure 8.5 depicts $k+1$ or two linear arrays consisting of m cells. The first array correspond to the level $l = 0$ and is similar to the array of Figure 8.4. The second array connected to each other via $k+2$ communication channels, one transferring the binary representation of text characters and the other $k+1$ transferring the bit-level results. Each row of the bit-level memory map R is allocated to a PE. Further, each cell performs a full step of the computation as shown in Figure 8.5 right, i.e. both mapping and OR/AND operations. The last communication channel which transfers bit-level results of level $l = 1$ travels from left to right in the same manner as the first two channels but at half speed.

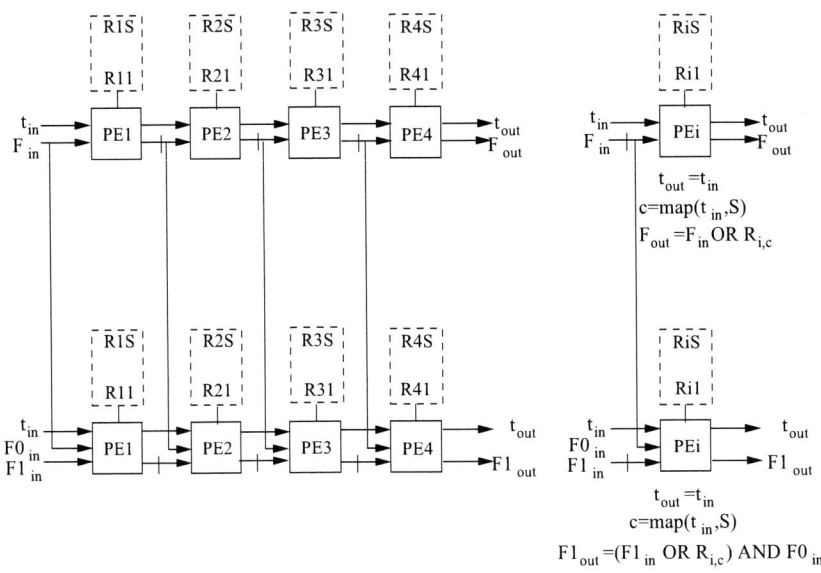

Figure 8.5. Processor array for string matching with mismatches.

Figure 8.6 shows $k + 1$ or two linear processor arrays of m cells. The first array corresponds to level $l = 0$ and it is similar to the array of Figure 8.4. We must note that a register F is allocated to each cell of the first array which stores the current value F as shown in Figure 8.4. The content of register F can be used as input to the corresponding cell of the second array. The second array connected to each other via $k + 3$ communication channels, one transferring the binary representation of text characters and the other $k + 2$ transferring the bit-level results. Each row of the R is allocated to a PE as in previous designs. Further, each cell performs computations as shown in Figure 8.6 right.

The total computation time for the three foregoing implementations is performed in $n + m - 1$ time steps. Further, the area required for the array of Figure 8.4 is m PEs, while the area for the arrays of Figures 8.5 and 8.6 is $(k+1)m$ PEs.

8.3. An Implementation of the Preprocessing Phase

In this section the implementation of the preprocessing phase, i.e. the construction of the bit-level memory map R is discussed. The aim is to use the same processor array architecture as in the previous section in order to produce the bit-level memory map with its elements allocated to the appropriate PEs. Furthermore, all the alternatives of flexible searching stated should be accommodated. For this reason each character p_i, $1 \leq i \leq m$, of the pattern p should be accompanied by some control information, denoting the presence of the don't care symbol, the complement symbol or the subrange symbol. Especially in the case of the subrange symbol two characters are required denoting the boundaries of the subrange. Therefore each augmented character p_i of the pattern consists of (i) at most two

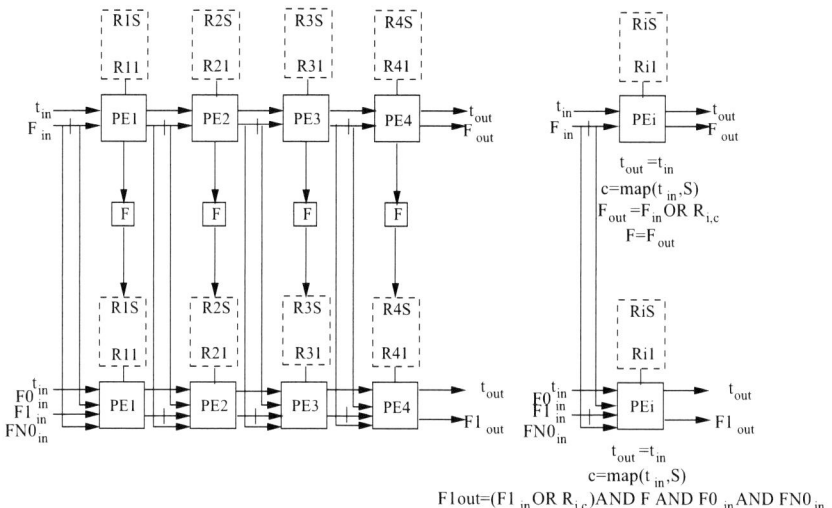

Figure 8.6. Processor array for string matching with differences.

bit strings which correspond to the binary codes of two valid characters belonging to Σ and (ii) necessary control information.

The operations that should be performed by the pre-processing phase are essentially writing operations to the appropriate row R_i, $1 \leq i \leq m$, of the bit-level memory map R that is allocated to the i-th PE. Since each memory location is a single bit the writing operation consists of setting this bit either to 0 or to 1. Using the example of the extended UNICODE character set, the following protocol is proposed in Table 8.1. Three control bits are specified: Bit 0 for setting a memory map bit either to 1 or to 0. Bits 1 and 2 denoting the presence of 0, 1 or 2 valid characters to be read.

Table 8.1. A preprocessing phase protocol

Functions	Ctrl 2	1	0	Chars 0/15	16/32
Clear (set to 0)	0	0	0	-	-
Don't care character (set to 1)	0	0	1	-	-
Simple character encoding	0	1	0	σ_1	-
Complement of a character	0	1	1	σ_1	-
Character subrange	1	1	0	σ_1	σ_2
Complement of character subrange	1	1	1	σ_1	σ_2

In order to implement the preprocessing phase protocol onto the linear processor array of Figure 8.2 the following assumptions are made. First, it is assumed that the text channel is used for transferring the character codes at a rate of a single character per transfer step, while the result channel is wide enough to carry the necessary control bits. Second, it is assumed that the alphabet Σ, its encoding and size $|\Sigma|$ are preloaded to the PEs. Third, the maximum length of the pattern is equal to the processor array size m and this is known to the system. Finally, the i-th PE should be aware of its position in the array, that is it should have a cell id equal to i, as numbered in Figure 8.2. The preprocessing phase can start by the reset signal through the control input and then each cell counts the loading steps that is j steps for the j-th PE. Therefore, each cell executes the following piece of pseudocode for the construction of the row of the bit-level memory map R:

$bit \leftarrow ctrl_0$
if $ctrl_1 = 0$ and $ctrl_2 = 0$ then $l_0 \leftarrow map(0, \Sigma), hi \leftarrow map(|\Sigma|, \Sigma)$
if $ctrl_1 = 1$ and $ctrl_2 = 0$ then $l_0 \leftarrow map(\sigma_1, \Sigma), hi \leftarrow map(\sigma_1, \Sigma)$
if $ctrl_1 = 1$ and $ctrl_2 = 1$ then $l_0 \leftarrow map(\sigma_1, \Sigma), hi \leftarrow map(\sigma_2, \Sigma)$
for $i = map(0, \Sigma)$ to $l_0 - 1$ do
    $R_i \leftarrow$ NOT bit
for $i = l_0$ to hi do
    $R_i \leftarrow bit$
for $i = hi + 1$ to $map(|\Sigma|, \Sigma)$ do
    $R_i \leftarrow$ NOT bit

There are $|\Sigma|$ writing operations to consecutive single bit locations in the row of the bit-level memory map M. The lowest address of the memory is denoted by $map(0, \Sigma)$ whereas the highest address is denoted by $map(|\Sigma|, \Sigma)$. All these memory locations are accessed once during the pre-processing phase, by means of three consecutive FOR loops. This allows to keep the computation time uniform for any type of preprocessing operation, that is $2m$ loading plus $|\Sigma|$ writing steps. The completion of the preprocessing phase enables the commencement of the searching phase. From the above description the preprocessing phase can be implemented in the same PE that performs the searching phase with the addition of limited programmable hardware. Therefore, the whole systolic algorithm can be performed onto a special purpose processor array.

Chapter 9

A Unified Array Processor Architecture

A number of commercial and research FPGAs implementations have been reported [29]. Examples are Splach-2 [36], Bioccelerator [21], Decypher [89], JBits [32], Hokiegene [76] and others [73, 100] are based on FPGAs. These solutions address the general problem of approximate string matching in the context of biological sequence data. Further, some researchers [82], [83] and [34] presented FPGAs implementations for simple string matching, regular expression matching and approximate string matching, respectively.

In the following, we describe the architecture of the programmable array processor and the architecture of the cell suitable for efficient execution of a class of flexible approximate string matching algorithms. There are a great number of approximate string matching algorithms in text retrieval, and programmability is required to accommodate these different algorithms within a single system. As new algorithms are developed, this architecture will be able to execute many of them without the need for redesigning the architecture. Further, the proposed architecture performs approximate string matching for simple and complex patterns, like don't care, complement and classes symbols.

9.1. Architecture of the Array

The architecture of the linear programmable array is shown in Figure 9.1. This structure was obtained as a result of applying the data dependence graph method [48] to the partitioned realization of a variety of approximate string matching algorithms presented in Section 6. In all those cases the method produced arrays with the same architecture, which led us to conclude that a class-specific array for the efficient execution of a class of algorithms was possible. The overall architecture is determined by the realization procedure presented in [48, 64], whereas the characteristics of the cells are determined by the requirements of the specific algorithms included in the class. As a result, the array matches well the capabilities of the method.

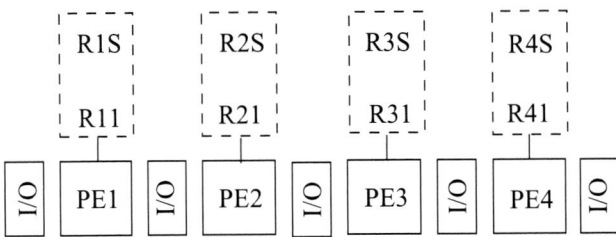

Figure 9.1. Linear programmable array processors for approximate string matching algorithms.

The array is one linear structure of m processing elements (PEs), as shown in Figure 9.1 for $m = 4$; five additional I/O interfaces, located at the intermediate of the cells, are used to simplify data communications that transfer the binary representation of text characters and the bit or byte-level results but do not perform any computations. The communication with the host is performed through the boundary PEs. This array executes all flexible approximate string matching algorithms, including the dynamic programming and the NFA algorithms. Communications are unidirectional, among neighbor cells and neighbor I/O modules. The operation of the linear array is controlled by a array controller. The array controller will be responsible for issuing signals to the array and for controlling communication with the host system.

In Figure 9.1 one character of the pattern and one row of the bit-level memory map R is preloaded into the PE of the array processor so that using the specialised addressing function $map(ch, \Sigma)$ produces a single bit per PE. For the implementation of the $map(ch, \Sigma)$ function we introduced a programmable hardware (decoder) in each cell. The other text string (or textbase) flows from left to right through the array. During each step, one elementary computation of any approximate string matching algorithm of the class is performed in each PE. The result is collected on the rightmost cell when the last character of the flowing string is output. If m is the length of the pattern and n is the length of the textbase, the computation of any algorithm in the class is performed in $m + n - 1$ steps on m PEs.

So far we have assumed an array processor equal in size of the pattern length. In practice, this rarely happens. Since the length of the patterns may vary, the computation of the dynamic programming algorithms must be partitioned on the fixed size array processor. The pattern length is usually larger than the array processor size. For sake of clarity we firstly assume a pattern of length m and an array processor of size r where m is a multiple of r. A possible solution is to split the computation into s passes: The first r characters of the pattern and r rows of the bit-level memory map R is loaded to the array processor. The entire textbase then crosses the array. In the next pass the following r characters of the pattern and rows of the bit-level memory maps are loaded into the array. The data stored previously is loaded together with the corresponding textbase and sent again through the array processor. The process is iterated until the end of the pattern is reached.

On the other hand, the computation of the NFA algorithms take the advantage of the intrinsic parallelism of the bit-level operations inside a word of the processor. In other words, each cell of the array will process a block of w characters of the pattern as the word size of the cell. That is, we pack w bit-level values in a single word and update them all in a single operation. This advantage of the intrinsic parallelism can not be used in the dynamic programming algorithms.

9.2. Architecture of the Cells

For the design of the programmable architecture, we examine the requirements (input, output, memory, registers and operations) of the cells of a variety of flexible approximate string matching algorithms and incorporated hardware within the cell to provide for fast execution of these algorithms while maintaining flexibility. Table 9.2 shows the requirements of the cells.

Therefore, the architecture of the processing element (PE) is shown in Figure 9.2. Each PE is connected to each other via three input and output communication channels when $k = 1$. One channel transferring the binary representation of text characters and another two transferring the bit-level (or byte-level) results. The input and output values in the lower part of the Figure 9.2 correspond to symbol names of the algorithms described in Section 6. More specifically, Table 9.1 shows the corresponding between the lower input/output values of Figure 9.2 and the symbol names of the alogrithms.

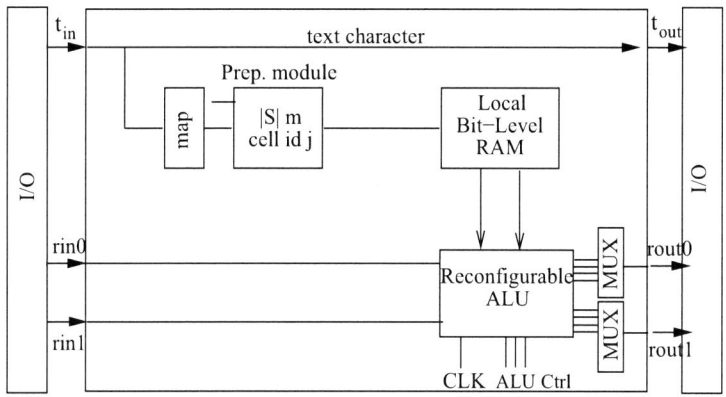

Figure 9.2. Cell specification for the programmable array processors.

The main components of the PE are the preprocessing module, character to address decoder, local bit-level random-access memory (RAM), arithmetic unit and multiplexers which are described as follows:

1. **Preprocessing Module**: This module is used to implement the preprocessing phase of several approximate string matching algorithms which we examined previously. Therefore, the preprocessing module has shown in Section 8.3.

2. **Character to Address Decoder**: From the text stream of Figure 9.2 it is observed that the binary code of the text character ch currently being transferred is used as input for the Character to Address Decoder. The decoder is essentially the programmable hardware implementation of the $map(ch, \Sigma)$ function. The output of the decoder is the address c of a bit-level memory location.

3. **Local Bit-Level Random Access Memory**: The local bit-level RAM keeps stored a row R_i, $1 \leq i \leq m$, of the bit-level memory map R. Therefore, the local memory size is $|\Sigma|$ bits and the address required to access such a memory is of $log|\Sigma|$ bits. Taking the example of the ASCII character set the local memory requirements are 64 bytes per cell. The result of the memory reading operation is two bits quantities $R_{i,c}$ and $M_{i,k}$ which in turn is one of the three operands of arithmetic unit.

4. **Arithmetic Unit**: The arithmetic unit consists of a arithmetic logic unit (ALU) and general purpose registers. The ALU is programmable and it consists of six small units, each unit implements all common logic functions and the standard arithmetic functions (addition or subtraction) of a string matching algorithm. In other words, the ALU supports partial reconfiguration because we will be reconfiguring only the ALU part of the circuit while leaving the rest of the modules untouched. When the user wants to execute any string matching algorithm, the ALU selects the suitable unit of the algorithm in each cell. Therefore, the unit is determined by three control bits which are used as input for decoder of the ALU. In ALU there is a decoder which is a hardware implementation of reconfiguration. The general purpose registers store intermediate results (such as D, Dv, DL, F, aux) and are usually directly connected to the data bus. This arithmetic unit is adequate for a set of flexible approximate string matching algorithms. However, an arithmetic unit with additional capabilities can be used to make the array processor suitable for other algorithms.

5. **Multiplexers**: Output multiplexers (MUX) of the ALU take care of selecting the proper communication channels that transfer the bit/byte-level results in the next cell when an algorithm is selected.

Table 9.1. Corresponding between the input/output values and the symbol names of algorithms

Algorithm	Input Output values					Registers				
	rin0	rin1	rout0	rout1	D	aux	Dv	F	DL
Dyn. Prog. - mismatches	$D_{i-1,j-1}$	-	$D_{i,j}$	$D_{i,j}$	-	-	-	-	-
Dyn. Prog. - differences	$D_{i-1,j}$	$D_{i-1,j-1}$	$D_{i,j}$	$D_{i,j}$	$D_{i,j-1}$	-	-	-	-
Dyn. Prog. - Myers	$Dh_{i-1,j}$	-	$Dh_{i,j}$	-	-	Dv	$Dv_{i,j-1}$	-	-
NFA - exact	F_i^0	-	F_i^0	-	-	-	-	-	-
NFA - mismatches	F_{i-1}^l	F_{i-1}^l	F_i^l	-	-	-	-	-	-
NFA - differences	F_{i-1}^l	F_{i-1}^{l-1}	F_i^l	-	-	-	-	F_i^l	-

Table 9.2. Requirements of the cells for several approximate string matching algorithms

Algorithm	Cells	Input	Output	Memory	Registers	Operations
Dyn. Prog. - mismatches	m	2	2	R	-	map/addition
Dyn. Prog. - differences	m	3	3	R	D	map/addition/min
Dyn. Prog. - Myers	m	2	2	R	Dv, aux	map/addition/substraction/min
NFA - exact	m	2	2	R	-	map/OR
NFA - mismatches	m	$k+2$	$k+1$	R	-	map/OR/AND
NFA - differences	m	$k+2$	$k+1$	R	F	map/OR/AND

Chapter 10

Comparison with Previous Hardware

The proposed hardware designs provide many advantages in relation to the other hardware approaches found in the literature. First of all, another systolic algorithm can be derived by applying a different transformation, as originally proposed in [25] for simple string matching. It allows for the concurrent pipelining of both text and search pattern through bi-directional data flow. This architecture has also been used for approximate string matching [36, 38, 56] by means of a parallelizing a dynamic programming algorithm but it exhibits very limited flexibility due to the encoding scheme used. A similar approach is pursued in [79, 80], using again a dynamic programming approach, but an improved encoding scheme and limited communication and control overheads. However, the bi-directional data flow imposes low processor utilization (approximately 50%) and increased computation time (approximately $2n$ steps instead of n steps, which is the case in the design proposed herein). An improvement in the time complexity of the bi-directional data flow architectures has been proposed in [60], which reduces the computation time to approximately n steps by partitioning the problem into two subproblems which are solved concurrently in the same array. This partitioning technique making the cell 100% efficient once the array is full. The architecture with uni-directional data flow presented herein achieves similar area and time complexity measures as the improved array without additional schedule complexity, i.e. partitioning or folding techniques. Also, the proposed architecture is closest to others [21, 32, 73, 76, 89, 100] since these architectures also use linear array of PEs on a reconfigurable platform. However, these designs make use of the limited flexibility since they does not allow for flexible patterns.

Further, our architecture implements an new category algorithms such as NFA algorithms. These algorithms require each cell to perform simple bit-level arithmetic and logical operations as opposed to the comparison operations used in other architectures [21, 32, 33, 36, 53, 54, 56, 60, 73, 76, 79, 80, 89]. The advantage of the bit-level arithmetic and logical operations is that are executed fast enough compared to the comparison operations.

Another important advantage of the proposed architecture presented herein is execution flexible approximate string matching algorithms as opposed to the previous architectures [21, 32, 33, 36, 53, 54, 56, 60, 73, 76, 79, 80, 89, 100] that perform simple approximate string matching. Therefore, we introduced the VLSI implementation of the encoding scheme, i.e. the introduction of the encoder and the bit-level memory modules. This encoding scheme enables the efficient implementation of the flexible approximate string matching algorithms in contrast to the limited flexibility of the encoding schemes used in the previous designs. Therefore, the programmability of our design makes it far more flexible than the other specialized systems.

Finally, the proposed design can be scaled with the pattern length while maintaining constant throughput but with a corresponding increase in latency. This design is also portable to other FPGAs.

Chapter 11

Conclusion

In this chapter we have presented implementations of flexible approximate string matching algorithms onto parallel architectures such as general-purpose parallel computers and special purpose parallel hardware. Further, we have presented the design and implementation of the proposed programmable and reconfigurable architecture for efficient execution of a class of approximate string matching algorithms. The software and hardware solutions derived in this chapter can speed up approximate string matching algorithms efficiently.

We expect that the hybrid computing can achieves supercomputer performance at low cost. Hybrid computing is the combination of the software and hardware solution within a parallel architecture, i.e. within the processors of a computer cluster programmable array processor boards are installed in order to accelerate compute intensive regular tasks. The driving force and motivation behind hybrid computing is the price/performance ratio. Using PC-cluster as in the Beowulf approach is currently the most efficient way to gain supercomputer power. Installing in addition massively programmable processor cards within each PC can further improve the cost/performance ratio significantly.

References

[1] A. Aho, M. Corasick, Efficient string matching: An aid to bibliographic search, *Communications of the ACM* **18**, 6, pp. 333-340 (1975).

[2] V.L. Arlazarov, E.A. Dinic, M.A. Kronrod, I.A. Faradzev, On economic construction of the transitive closure of a directed graph, *Dokl. Akad. Nauk SSSR* **194**, pp. 487-488 (1970) (in Russian). English translation in *Soviet Math. Dokl.* **11**, pp. 1209-1210 (1975).

[3] R.S. Boyer, J.S. Moore, A fast string searching algorithm, *Communications of the ACM* **20**, 10, pp. 762-772 (1977).

[4] R. Baeza-Yates, Text Retrieval: Theory and Practice, in Proc. of the 12th IFIP World Computer Congress, pp. 465-476, (Madrid, Spain), North-Holland (1992).

[5] R. Baeza-Yates, G.H. Gonnet, A new approach to text searching, *Communications of the ACM* **35**, 10, pp. 74-82 (1992).

[6] R.A. Baeza Yates, G. Gonnet, Fast string matching with mismatches, *Information and Computation* **108**, 2, pp. 187-199 (1994).

[7] R. Baeza-Yates, A unified view of string matching algorithms, in *Proc. of SOFSEM'96: Theory and Practice of Informatics* (1996).

[8] R.A. Baeza-Yates, G. Navarro, A fast heuristic for approximate string matching, in: *Proc. of the 3rd South American Workshop on String Processing*, (Carleton University Press), pp. 47-63 (1996).

[9] R.A. Baeza-Yates, G. Navarro, A faster algorithm for approximate string matching, in: *Proc. of the 7th Annual Symposium on Combinatorial Pattern Matching*, No. **1075** (Springer-Verlag, Berlin), pp. 1-23 (1996).

[10] R.A. Baeza-Yates, G. Navarro, A faster algorithm for approximate string matching, *Algorithmica* **23**, 2, pp. 127-158 (1999).

[11] R.A. Baeza-Yates, C.H. Perleberg, Fast and practical approximate string matching, *Information Processing Letters* **59**, 1, pp. 21-27 (1996).

[12] A.A. Bertossi, F. Logi, Parallel string matching with variable length don't cares, *Journal of Parallel and Distributed Computing* **22**, 2, pp. 229-234 (1994).

[13] R.D. Bjornson, A.H. Sherman, S.B. Weston et. al, TurboBLAST: A Parallel Implementation of BLAST Built on the TurboHub, *International Parallel and Distributed Processing Symposium: IPDPS Workshops,* p.0183 (2002).

[14] A. Boukerche, A.C.M.A. Melo, M. Ayala-Rincon, T.M. Santana,, Parallel strategies for local biological sequence alignment in a cluster of workstations, in *Proc. of the 19th IEEE International Parallel and Distributed Processing Symposium* (2005).

[15] R. Buyya, *High Performance Cluster Computing: Architectures and Systems,* volume 1, Prentice Hall (1999).

[16] R. Buyya, *High Performance Cluster Computing: Programming and Applications,* volume 2, Prentice Hall (1999).

[17] W.I. Chang, J. Lampe, Theoretical and Empirical Comparisons of approximate string matching algorithms. In: *Proc. of the 3rd Annual Symposium on Combinatorial Pattern Matching,* No. **664** (Springer-Verlag, Berlin), pp. 175-184 (1992).

[18] J. Cheetham, F. Dehne, S. Pitre, A. Rau-Chaplin, P.J. Taillon, Parallel CLUSTAL W For PC Clusters, *International Conference on Computational Sciences and Its Applications* (2003).

[19] H.D. Cheng, K.S. Fu, VLSI architectures for string matching and pattern matching, *Pattern Recognition* **20**, 1, pp. 125-141 (1987).

[20] B. Commentz-Walter, A string matching algorithm fast on the average, in: *Proc. of the 611 International Colloquium on Automata, Languages and Programming,* No. **71** (Springer-Verlag, Berlin), pp. 118-132 (1979).

[21] Compugen Ltd. Bioccelerator information package, Obtained from compugen@datasrv.co.il (1994).

[22] M. Crochemore, A. Czumaj, L. Gasieniec, S. Jarominek, T. Lecroq, W. Plandowski, W. Rytter, Speeding Up Two String Matching Algorithms, *Algorithmica* **12**, 4-5, pp. 247-267 (1994).

[23] M. Crochemore, W. Rytter, *Text Algorithms,* Oxford University Press (1994).

[24] J. Cringean, R. England, G. Manson, P. Willett, Network design for the implementation of text searching using a multicomputer, *Information Processing Management* **27**, 4, pp. 265-283 (1991).

[25] M.J. Foster, H.T. Kung, The design of special purpose VLSI chip, *IEEE Computer* **13**, pp. 26-40 (1980).

[26] Z. Galil, A constant time optimal parallel string matching algorithm, *Journal of the ACM* **42**, 4, pp 908-918 (1995).

[27] Z. Galil, K. Park, An improved algorithm for approximate string matching, *SIAM Journal of Computing* **19**, 6, pp. 989-999 (1990).

[28] A. Geist, A. Beguelin, J. Dongarra, W. Jiang, R. Manchek, V. Sunderam, PVM: Parallel Virtual Machine, *A Users Guide and Tutorial for Networked Parallel Computing*, Massachusetts: The MIT Press (1994).

[29] M. Gokhale, P. S. Graham, *Reconfigurable Computing: Accelerating Computation with Field-Programmable Gate Arrays,* Springer-Verlag, 1st edition (2006).

[30] W. Gropp, E. Lusk, A. Skjellum, *Using MPI: Portable Parallel Programming with the Message Passing Interface*, MIT Press, Gambridge, Massachusetts (1994).

[31] R. Grossi, F. Luccio, Simple and efficient string matching with k mismatches, *Information Processing Letters* **33**, 3, pp. 113-120 (1989).

[32] S. Guccione, E. Keller, Gene matching using JBits, in Proc. of the 12th International Conference on Field Programmable Logic and Applications, *LNCS* **2438**, pp. 1168-1171 (2002).

[33] P. Guerdoux-Jamet, D. Lavenier, C. Wagner, P. Quinton, Design and implementation of a parallel architecture for biological sequence comparison, in *Proc. of Euro-Par* (1996).

[34] M.C. Herbordt, T. VanCourt, Y. Gu, J. Model, B. Sukhwani, Single Pass Approximate String Matching on FPGAs, in *Proc. of IEEE Symposium on Field-Programmable Custom Computing Machines* (2006).

[35] J. D. Hirschberg, R. Hughey, K. Karplus: Kestrel: A programmable array for sequence analysis, in *Proc. of International Conference Application-Specific Systems, Architectures and Processors*, pp. 25-34, IEEE CS (1996).

[36] D.T. Hoang, Searching genetic databases on Splash 2, in *Proc. of IEEE Workshop on FPGAs for Custom Computing Machines,* pp. 185-191 (1993).

[37] R.N. Horspool, Practical fast searching in strings, *Software-Practice and Experience* **10**, 6, pp. 501-506 (1980).

[38] R. Hughey, D.P. Lopresti, Architecture of a programmable systolic array, in *Proc. of International Conference Systolic Arrays,* pp. 41-50 (1988).

[39] R. Hughey, Parallel sequence comparison and alignment, in *Proc. of the International Conference Application-Specific Array Processors* (1995).

[40] R. Hughey, *Parallel hardware for sequence comparison and alignment, CABIOS* **12**, 6, pp. 473-479 (1996).

[41] C. Janaki, R.R. Joshi, Accelerating comparative genomics using parallel computing, *Silicon Biology* **3**, 0036 (2003).

[42] Y. Jiang, A.H. Wright, O(k) parallel algorithms for approximate string matching, *Neural, Parallel and Scientific Computations* **1** pp. 443-452 (1993).

[43] P. Jokinen, J. Tarhio, E. Ukkonen, A comparison of approximate string matching algorithms, *Software-Practice and Experience* **26**, 12, pp. 1439-1458 (1996).

[44] A. Julich, Implementations of BLAST for parallel computers, *Computers Applications Bioscience* **11**, 1, pp. 3-6 (1995).

[45] D.E. Knuth, J.H. Morris, V.R. Pratt, Fast pattern matching in strings, *SIAM Journal on Computing* **6**, 2, pp. 323-350 (1977).

[46] Kuo-Bin Li, ClustalW-MPI: ClustalW analysis using distributed and parallel computing, *Bioinformatics* (2003).

[47] S. Kurtz, Approximate string searching under weighted edit distance, in: *Proc. of the 3rd South American Workshop on String Processing* (Carleton University Press), pp. 156-170 (1996).

[48] S.Y. Kung, VLSI Array Processors, Prentice-Hall (1988).

[49] G.M. Landau, U. Vishkin, Efficient string matching with k mismatches, *Theoretical Computer Science* **43**, 2-3, pp. 239-249 (1986).

[50] G.M. Landau, U. Vishkin, Fast string matching with k diffrrences, *Journal of Computer and System Sciences* **37**, 1, pp. 63-78 (1988).

[51] G.M. Landau, U. Vishkin, Fast parallel and serial approximate string matching, *Journal of Algorithms* **10**, 2, pp. 157-169 (1989).

[52] D. Lavenier, J.L. Pacherie, Parallel processing for scanning genomic data-bases, in: *Proceedings PARCO '97*, pp. 81-88 (1997).

[53] D. Lavenier, *SAMBA: Systolic accelerators for molecular biological applications*, Technical Report 988 IRISA 35042 Rennes Cedex, France (1996).

[54] D. Lavenier, Speeding up genome computations with a systolic accelerator, *SIAM News* **31**, 8, pp. 6-7 (1998).

[55] T. Lecroq, A variation on the Boyer-Moore algorithm, *Theoretical Computer Science* **92**, 1, pp. 119-144 (1992).

[56] D.P. Lopresti, P-NAC: A systolic array for comparing nucleic acid sequences, *Computer* **20**, 1, pp. 98-99 (1987).

[57] R.J. Lipton, D.P. Lopresti, A systolic array for rapid string comparison, in *Proc. Chapel Hill Conference on VLSI*, pp. 363-376 (1985).

[58] K.G. Margaritis, D.J. Evans, A VLSI processor array for flexible string matching, *Parallel Algorithms and Applications* **11**, pp. 45-60 (1997).

[59] W.J. Masek, M.S. Paterson, A faster algorithm computing string edit distances, *Journal of Computer and System Sciences* **20**, pp. 18-31 (1980).

[60] G.M. Megson, Efficient systolic string matching, *Electronic Letters* **26**, 24, pp. 2040-2042 (1990).

[61] P.D. Michailidis, K.G. Margaritis, *String Matching Algorithms, Technical Report*, Department of Ap. Informatics, University of Macedonia, in Greek (1999).

[62] P.D. Michailidis, K.G. Margaritis, On-line string matching algorithms: Survey and experimental results, *International Journal of Computer Mathematics* **76**, pp. 411-434 (2001).

[63] P.D. Michailidis, K.G. Margaritis, On-line approximate string searching algorithms: survey and experimental results, *International Journal of Computer Mathematics* **79**, 8, pp. 867-888 (2002).

[64] P.D. Michailidis, *Parallel and Distributed Implementations for Approximate String Matching*, Ph.D. Thesis, Department of Applied Informatics, University of Macedonia (2004) (in Greek).

[65] G. Myers, A fast bit-vector algorithm for approximate pattern matching based on dynamic programming, in: *Proc. of the 9h Annual Symposium on Combinatorial Pattern Matching*, No. **1448** (Springer-Verlag, Berlin), pp. 1-13 (1998).

[66] G. Myers, A fast bit-vector algorithm for approximate pattern matching based on dynamic programming, *Journal of the Association for Computing Machinery* **46**, 3, pp. 395-415 (1999).

[67] G. Navarro, Mutiple approximate string matching by counting, in: *Proc. of the 4th South American Workshop on String Processing* (Carleton University Press), pp. 125?139 (1997).

[68] G. Navarro, A partial deterministic automaton for approximate string matching. In: *Proc. of the 4th South American Workshop on String Processing* (Carleton University Press), pp. 112-124 (1997).

[69] G. Navarro, M. Raffinot, A Bit-Parallel Approach to Suffix Automata: Fast Extended String Matching, in *Proc. of the 9th Annual Symposium on Combinatorial Pattern Matching* **1448**, pp. 14-33, Springer-Verlag, Berlin (1998).

[70] G. Navarro, R. Raffinot, Fast and flexible string matching by combining bit- parallelism and suffix automata, *ACM Journal of Experimental Algorithmics* **5**, 4 (2000).

[71] G. Navarro, A guided tour to approximate string matching, *ACM Computing Surveys* **33**, 1 pp. 31-88 (2001).

[72] B. Nichols, D. Buttlar, J.P. Farrell, Pthreads Programming, OReilly (1996).

[73] T. Oliver, B. Schmidt, D. Maskell, *Hyper Customized Processors for Bio-Sequence Database Scanning on FPGAs*, in Proc. of FPGA05 (2005).

[74] P.S. Pacheco, *Parallel Programming with MPI*, San Francisco, CA, Morgan Kaufmann (1997).

[75] G. Pfister, *In Search of Clusters*, Prentice Hall PTR, NJ, 2nd Edition, NJ (1998).

[76] K. Puttegowda et al, A run-time reconfigurable system for gene-sequence searching, in *Proc. of 16th International Conference on VLSI Design*, pp. 561-566 (2003).

[77] N. Ranganathan, R. Sastry, VLSI architectures for pattern matching, *International Journal of Pattern Recognition and Artificial Intelligence* **8**, 4, pp. 815-843 (1994).

[78] A.J. Ropelewski, H.B. Nicholas, D.W. Deerfield, Implementation of genetic sequence alignment programs on supercomputers, *Journal of Supercomputing* **11**, pp. 237-253 (1997).

[79] R. Sastry, N. Ranganathan, K. Remedios, CASM: A VLSI chip for approximate string matching, *IEEE Transactions on Pattern Analysis and Machine Intelligence*, **17**, 8, pp. 824-830 (1995).

[80] R. Sastry, N. Ranganathan, A systolic array for approximate string matching, in *Proc. of International Conference Computer Design*, pp. 402-405 (1993).

[81] P.H. Sellers, The Theory and Computation of Evolutionary Distance: Pattern Recognition, *Journal of Algorithms* **1**, 4, pp. 359-373 (1980).

[82] R. Sidhu, A. Mei, V.K. Prasanna, String matching on multicontext FPGAs using self-reconfiguration, in *Proc. of International Symposium on Field-Programmable Gate Arrays* (1999).

[83] R. Sidhu, V.K. Prasanna, Fast regular expression matching using FPGAs, in *Proc. of IEEE Symposium on Field-Programmable Custom Computing Machines* (2001).

[84] R.K. Singh et al, BIOSCAN: A network sharable computational resource for searching biosequence databases, *CABIOS* **12**, 3, pp. 191-196 (1996).

[85] P. Smith, Experiments with a very fast substring search algorithm, *Software-Practice and Experience* **21**, 10, pp. 1065-1074 (1991).

[86] M. Snir, S. Otto, S. Huss-Lederman, D. W. Walker, J. Dongarra, *MPI: The complete reference*, Massachusetts: The MIT Press (1996).

[87] D. Sunday, A very fast substring search algorithm, *Communications of the ACM* **33**, 8, pp.132-142 (1990).

[88] J. Tarhio, E. Ukkonen, Approximate boyer-moore string matching, *SIAM Journal on Computing* **22**, 2, pp. 243-260 (1993).

[89] TimeLogic Corporation, *Decypher bioinformatics acceleration solution*, http://www.timelogic.com/decypher_intro.html (2002).

[90] O. Trelles, On the Parallelization of Bioinformatic Applications, *Briefings in Bioinformatics* **2** (2001).

[91] E. Ukkonen, Algorithms for approximate string matching, *Information and Control* **4**, 1-3, pp. 100-118 (1985).

[92] E. Ukkonen, Finding approximate patterns in strings, *Journal of Algorithms* **4**, 1-3, pp. 132-137 (1985).

[93] B. Ukkonen, D. Wood, Approximate string matching with suffix automata, *Algorithmica* **10**, 5, pp. 353-364 (1993).

[94] R.A. Wagner, M.J. Fischer, The string to string correction problem, *Journal of the Association for Computing Machinery* **21**, 1, pp. 168-173 (1974).

[95] B. Wilkinson, M. Allen, *Parallel Programming: Techniques and Applications using Networked Workstations and Parallel Computers*, Prentice Hall, Englewood Cliffs, NJ, 2n editionn (2005).

[96] S. Wu, U. Manber, Fast text searching allowing errors, *Communications of the ACM* **35**, 10, pp. 83-91 (1992).

[97] S. Wu, U. Manber, G. Myers, A subquadratic algorithm for approximate limited expression matching, *Algorithmica* **15**, 1, pp. 50-67 (1996).

[98] T.K. Yap, O. Frieder, R.L. Martino, *High Performance Computational Methods for Biological Sequence Analysis*, Kluwer Academic Publishers (1996).

[99] T. K. Yap, O. Frieder, R. L. Martino, Parallel computation in biological sequence analysis, *IEEE Transactions Parallel Distributed Systems* **9**, 3, pp. 283-293 (1998).

[100] C.W. Yu, K.H. Kwong, K.H. Lee, P.H.W. Leong, A Smith-Waterman Systolic Cell, in Proc. of the 13th International Conference on Field-Programmable Logic and Applications, Springer-Verlag *LNCS* **2778**, pp. 375-384 (2003).

Index

A

accelerator, 65
access, 16, 17, 19, 54
ACM, 61, 63, 66, 67, 68
adaptation, 10, 11
aid, 61
algorithm, 2, 3, 4, 5, 6, 7, 8, 9, 10, 11, 12, 13, 16, 22, 25, 26, 29, 30, 32, 35, 41, 47, 50, 52, 54, 57, 61, 63, 65, 66, 67, 68
alphabets, 6
alternative, 15, 16, 17, 19, 26
alternatives, 12, 36, 48
application, 16, 18, 20, 21
arithmetic, 5, 53, 54, 57
assignment, 22, 46
assumptions, 25, 27, 49
asynchronous, 20, 22
attention, 10
automata, 3, 5, 7, 29, 66, 68
availability, 19

B

bandwidth, 17
behavior, 20
benefits, 17, 18
bioinformatics, vii, 67
biological, 25, 51, 63, 64, 65, 68
biology, 2
blocks, 11
bottleneck, 21
bypass, 19

C

capacity, 27
category a, 57
cell, vii, 22, 31, 45, 46, 47, 48, 50, 51, 52, 53, 54, 57
channels, 45, 46, 47, 48, 53, 54
classes, 51
classical, 3, 7, 8, 10, 29
classified, 20
closure, 61
clusters, 16, 17, 19, 20
codes, 49
coding, 3, 7
commercial, 51
communication, vii, 16, 18, 19, 20, 21, 22, 26, 27, 45, 46, 47, 48, 52, 53, 54, 57
communication overhead, vii, 19, 26, 27
communities, 2
complement, vii, 2, 37, 48, 51
complexity, 3, 4, 5, 8, 11, 13, 22, 37, 45, 57
components, 15, 17, 18, 53
computation, 20, 21, 22, 26, 30, 31, 32, 36, 40, 45, 46, 47, 48, 50, 52, 57, 68
computer, vii, 1, 2, 5, 12, 15, 16, 18, 59
computer science, 1
computers, 15, 16, 17, 18, 19, 65, 68
computing, vii, 1, 2, 4, 10, 15, 17, 18, 19, 20, 31, 59, 64, 65
concurrency, 16
configuration, 16
Congress, 61
construction, 7, 29, 48, 50, 61
consumers, 21
control, 20, 48, 49, 50, 54, 57
controlled, 52
cost-effective, 15, 18
costs, 1, 5, 15
cycles, 21

D

data communication, 19, 52
data structure, 3, 4
definition, 31
degree, 41, 42, 44
delta, 31, 45, 46
density, 16
deterministic, 5, 7, 10, 66
DFA, 10
distributed memory, 16, 25
distributed memory computer, 16
distribution, 11, 21, 27
DSM, 19

E

economic, 61
electronic, 19
encoding, 49, 57, 58
engineering, 20
England, 63
English, 9, 12, 13, 61
environment, 18, 20, 26
execution, 2, 19, 21, 51, 53, 57, 59

F

fabrication, 15
farm, 21, 26
flexibility, 5, 16, 53, 57, 58
flow, 46, 57
folding, 57
Fortran, 20
FPGA, 2, 16
FPGAs, 16, 51, 58, 64, 66, 67
France, 65

G

generalization, 1
generation, 21
genetic, 64, 67
genome, 65
genomic, 65
genomics, 64
graph, 21, 22, 39, 40, 41, 42, 43, 44, 45, 47, 51, 61

H

heterogeneous, vii, 2, 25, 26, 27
heuristic, 4, 61
high-level, 20
homogenous, 25
host, 52
hybrid, vii, 27, 59

I

IBM, 16
id, 50, 53
illusion, 19
implementation, vii, 2, 7, 20, 22, 25, 26, 27, 29, 45, 48, 51, 52, 53, 54, 57, 58, 59, 63, 64
industry, 20
information retrieval, vii, 2
insertion, 1, 36
Intel, 16
interaction, 21, 22
interface, 19, 20

intrinsic, 5, 52, 53

L

LAN, 17, 18
language, 17
large-scale, 17
latency, 17, 58
laws, 15
limitation, 15
linear, vii, 3, 5, 21, 22, 23, 45, 46, 47, 48, 49, 51, 52, 57
literature, 20, 57
load imbalance, vii, 26, 27
location, 49, 53

M

Macedonia, 66
machines, 17, 20
management, 19
mapping, 21, 22, 30, 45, 46, 47
market, 17
Massachusetts, 64, 67
matrix, 29, 30, 31, 32, 36, 39, 41
measures, 57
memory, 10, 16, 17, 18, 20, 21, 25, 26, 29, 30, 39, 45, 47, 48, 49, 50, 52, 53, 54, 57
message passing, vii, 17, 19, 20
messages, 19
middleware, 19
MIT, 64, 67
models, 20
modules, 52, 54, 57
motivation, 59
MPI, vii, 2, 17, 19, 20, 25, 26, 27, 64, 66, 67
multidimensional, 21

N

network, 16, 17, 19, 22, 27, 67
networking, 17
NICs, 19
nodes, 18, 19, 22, 39, 42, 43, 44
nucleic acid, 65

O

online, 3
on-line, 2, 3, 7
operating system, 18, 19

P

packets, 19, 20
paper, 4, 8
parallel algorithm, 2, 64
parallel computers, 15, 17, 25, 59, 65
parallel implementation, vii, 27
parallel methods, vii, 2
parallel platform, 17
parallel processing, 16, 17, 19
parallel simulation, 13
parallelism, 5, 16, 21, 22, 29, 52, 53
parallelization, 12, 13, 26
parameter, 26
partition, 10, 11, 25
PCs, 17, 18, 20
performance, vii, 2, 4, 8, 15, 17, 18, 26, 30, 59
personal, 17
pipelines, 21
pipelining, 16, 22, 57
platforms, 17
point-to-point, 20
poor, 26
power, 15, 59
preprocessing, vii, 3, 4, 5, 7, 8, 10, 11, 12, 13, 29, 30, 32, 37, 48, 49, 50, 53
producers, 21
programmability, 51, 58
programming, vii, 10, 11, 12, 13, 19, 20, 25, 29, 30, 31, 32, 36, 39, 41, 44, 45, 52, 53, 57, 66
property, 31
proposition, 17
protocol, 49
protocols, 17, 19
prototype, 17
PVM, 19, 20, 64

R

random, 8
range, 15, 35, 36
reading, 42, 45, 54
recurrence, 31
reduction, 22
regular, 22, 51, 59, 67
research, 17, 51
researchers, 10, 20, 39, 51
resources, 20
routines, 20
Russian, 61

S

scheduling, 22
scientific, 20

search, 3, 5, 10, 11, 12, 26, 29, 30, 31, 32, 36, 37, 57, 61, 67
searches, 2
searching, vii, 1, 2, 4, 5, 7, 8, 10, 11, 12, 29, 32, 34, 35, 39, 45, 48, 50, 61, 64, 65, 66, 67
SEL, 10
selecting, 54
semantics, 20
services, 18
shape, 21
sharing, 15
signals, 52
simulation, 32, 33, 34, 35
software, 19, 20, 59
solutions, 2, 17, 51, 59
South America, 61, 65, 66
Spain, 61
speed, 2, 15, 27, 45, 46, 48, 59
speed of light, 15
SSI, 19
standardization, 17
standards, 17
strategies, 25, 63
string matching, vii, 2, 3, 5, 6, 9, 11, 12, 13, 15, 16, 25, 26, 27, 29, 30, 32, 33, 34, 35, 36, 39, 40, 41, 42, 43, 44, 46, 47, 48, 49, 51, 52, 53, 54, 55, 57, 58, 59, 61, 63, 64, 65, 66, 67, 68
string matching problem, vii, 2, 3, 29
substitution, 1, 34, 35, 36
subtraction, 54
Sunday, 4, 67
supercomputers, 67
supply, 16
symbols, 36, 51
synchronization, 21
syntax, 20
systematic, 22
systems, 15, 16, 17, 18, 20, 58

T

technology, 16
text searching, vii, 12, 61, 63, 68
theoretical, 2
thermodynamic, 15
time, vii, 2, 3, 4, 5, 8, 10, 11, 12, 13, 22, 26, 27, 32, 41, 42, 43, 44, 45, 46, 48, 50, 57, 63
timing, 39, 40, 41, 42, 43, 44
tin, 46, 47, 48, 49, 53
torus, 16
trade, 10
transfer, 49, 52, 54
transformation, 57
transition, 10, 34
transitions, 11, 34
translation, 61
trees, 21
trend, 11, 17

triggers, 21
two-way, 21

U

uniform, 27, 50
users, 18

V

values, 4, 10, 30, 31, 32, 33, 34, 35, 36, 41, 42, 46, 52, 53, 55
variable, 61
variables, 20

variation, 10, 36, 65
vector, 29, 32, 36, 42, 44
visible, 16

W

waiting times, 21
wealth, 36
WM, 12
workers, 21, 26, 27
workstation, 17, 18, 25, 26, 27
writing, 16, 20, 49, 50